McDougal Littell
Algebra 1
Concepts and Skills

Larson   Boswell   Kanold   Stiff

CHAPTER 1 # Resource Book

The Resource Book contains the wide variety of black-line masters available for Chapter 1. The blacklines are organized by lesson. Included are support materials for the teacher as well as practice, activities, applications, and assessment resources.

McDougal Littell
A HOUGHTON MIFFLIN COMPANY
Evanston, Illinois • Boston • Dallas

Contributing Authors

The authors wish to thank the following individuals for their contributions to the Chapter 1 Resource Book.

Rita Browning
Linda E. Byrom
José Castro
Rebecca S. Glus
Christine A. Hoover
Carolyn Huzinec
Karen Ostaffe
Jessica Pflueger
Barbara L. Power
James G. Rutkowski
Michelle Strager

2004 Impression

ISBN-13: 978-0-618-07851-6   ISBN-10: 0-618-07851-7

13 14 -HWI- 09 08

Contents

Contents

Contents

Descriptions of Resources

This Chapter Resource Book is organized by lessons within the chapter in order to make your planning easier. The following materials are provided:

Tips for New Teachers These teaching notes provide both new and experienced teachers with useful teaching tips for each lesson, including tips about common errors and inclusion.

Parent Guide for Student Success This guide helps parents contribute to student success by providing an overview of the chapter along with questions and activities for parents and students to work on together.

Prerequisite Skills Review Worked-out examples are provided to review the prerequisite skills highlighted on the Study Guide page at the beginning of the chapter. Additional practice is included with each worked-out example.

Strategies for Reading Mathematics The first page teaches reading strategies to be applied to the current chapter and to later chapters. The second page is a visual glossary of key vocabulary.

Lesson Plans and Lesson Plans for Block Scheduling This planning template helps teachers select the materials they will use to teach each lesson from among the variety of materials available for the lesson. The block-scheduling version provides additional information about pacing.

Warm-Up Exercises and Daily Homework Quiz The warm-ups cover prerequisite skills that help prepare students for a given lesson. The quiz assesses students on the content of the previous lesson. (Transparencies also available)

Activity Support Masters These blackline masters make it easier for students to record their work on selected activities in the Student Edition.

Alternative Lesson Openers An engaging alternative for starting each lesson is provided from among these four types: *Application, Activity, Graphing Calculator,* or *Visual Approach.* (Color transparencies also available)

Graphing Calculator Activities with Keystrokes Keystrokes for four models of calculators are provided for each Technology Activity in the Student Edition, along with alternative Graphing Calculator Activities to begin selected lessons.

Practice A and B These exercises offer additional practice for the material in each lesson, including application problems. There are two levels of practice for each lesson: A (transitional) and B (average).

Contents

Reteaching with Practice These two pages provide additional instruction, worked-out examples, and practice exercises covering the key concepts and vocabulary in each lesson.

Quick Catch-Up for Absent Students This handy form makes it easy for teachers to let students who have been absent know what to do for homework and which activities or examples were covered in class.

Learning Activities These enrichment activities apply the math taught in the lesson in an interesting way that lends itself to group work.

Interdisciplinary Applications/Real-Life Applications Students apply the mathematics covered in each lesson to solve an interesting interdisciplinary or real-life problem.

Challenge: Skills and Applications Teachers can use these exercises to enrich or extend each lesson.

Quizzes The quizzes can be used to assess student progress on two or three lessons.

Chapter Review Games and Activities This worksheet offers fun practice at the end of the chapter and provides an alternative way to review the chapter content in preparation for the Chapter Test.

Chapter Tests A and B These are tests that cover the most important skills taught in the chapter. There are two levels of test: A (transitional) and B (average).

SAT/ACT Chapter Test This test also covers the most important skills taught in the chapter, but questions are in multiple-choice and quantitative-comparison format. (See *Alternative Assessment* for multi-step problems.)

Alternative Assessment with Rubrics and Math Journal A journal exercise has students write about the mathematics in the chapter. A multi-step problem has students apply a variety of skills from the chapter and explain their reasoning. Solutions and a 4-point rubric are included.

Project with Rubric The project allows students to delve more deeply into a problem that applies the mathematics of the chapter. Teacher's notes and a 4-point rubric are included.

Cumulative Review These practice pages help students maintain skills from the current chapter and preceding chapters.

Tips for New Teachers

For use with Chapter 1

LESSON 1.1

TEACHING TIP To help students understand the idea of variable, evaluate some "fill in" problems like $8 + \underline{}$. Students must evaluate it by filling in the gap with a number that they choose (students must state beforehand what number they will use to "fill in" the gap). Once the students are comfortable with this idea, you can rewrite the "fill in" problem as the variable expression $8 + x$. Now students can see that evaluating a variable expression is just a matter of replacing the variable by its given value. To reinforce the idea that a variable can represent one or more numbers, evaluate the same expression with different values of the variable such as $x = 2$ or $x = 9$.

COMMON ERROR Some students may confuse the area formula with the perimeter formula. A very common mistake is to add two sides of a rectangle to find its perimeter. Make a drawing of a rectangle and label its sides with the corresponding lengths. Use the drawing to show the perimeter as the distance around the shape and the area as the space inside of the shape. You might also discuss the units to use to measure area and perimeter.

TEACHING TIP Help students to complete unit analysis by rewriting expressions involving fractions in a horizontal manner. For instance, the quotient $\dfrac{\text{mi}}{\text{mi/h}}$ can be written as $\text{mi} \div \dfrac{\text{mi}}{\text{h}}$. You can now use properties of fractions to rewrite this expression as $\text{mi} \cdot \dfrac{\text{h}}{\text{mi}}$.

LESSON 1.2

COMMON ERROR Students may multiply the base times the exponent of a power to find its value. Students making this error may interpret 5^3 as $5 \cdot 3 = 15$. Emphasize that the exponent of a power is a "counter" that indicates how many times to repeat the base. In other words, 5^3 means that we need to multiply three 5's: $5 \cdot 5 \cdot 5$. When we rewrite the power notation as a repeated product of the base, the exponents disappear.

INCLUSION Students with limited English proficiency should write down the meanings of the expressions "squared" and "cubed" with appropriate examples.

COMMON ERROR When working with problems such as Example 5 on page 11, some students will find the correct answer for the volume and then cube it. These students do not understand volume units such as ft^3. They believe that the exponent used to abbreviate cubic feet affects the numeric value of the answer. Review the appropriate way to write units of length, area, and volume.

TEACHING TIP To practice order of operations, give students a set of similar problems to solve, such as $2^2 + 3^2$, $(2^2 + 3)^2$, $(2 + 3)^2$, $((2 + 3)^2)^2$.

LESSON 1.3

TEACHING TIP You can start the lesson by asking students to evaluate a simple expression such as $12 - 6 \div 2$. Students who are familiar with order of operations will obtain the correct answer, but other students will get incorrect responses. Ask students to explain how they got their answers to show that they did not make any computational errors: they got different responses because they followed a different order of operations.

COMMON ERROR Some students may already be familiar with "Please Excuse My Dear Aunt Sally." This helpful trick to remember order of operations could lead students to believe that multiplication should always be performed before division and addition should always precede subtraction. To avoid this misconception, have students write the phrase on different lines:

Please

Excuse

My Dear

Aunt Sally

Operations written on the same line have the same priority and must be performed using the "left-to-right" rule.

Tips for New Teachers

For use with Chapter 1

LESSON 1.4

INCLUSION The difference between $>$ and $\geq$ is not immediately clear for some students. You can use examples such as $2x - 1 > 7$ and $2x - 1 \geq 7$ to discuss whether 4 is a possible solution and to show the difference between these inequalities.

INCLUSION Expressions such as "no more than", "no less than", "at least", and "at most" should be discussed carefully.

LESSON 1.5

INCLUSION Students with limited English may struggle with the reading in this lesson. These students might need a shorter homework assignment. Have students keep a list of "key words" in their notes. This list should contain words that mean addition, subtraction, multiplication, and division, grouped by operation. You can start the class by writing those four words on the board and asking students to come up with other words that mean the same thing. Students should add words to their list as they complete their classwork or homework. Make sure to probe for understanding with these students: ask them to tell you what the problem says using their own words. Do not give up: this is a topic that students will master over time, not the first day.

LESSON 1.6

TEACHING TIP You can start the lesson with a scenario such as a clothing purchase. If the students wanted to buy three pairs of jeans at $40.00 each, how much is the total sale?

LESSON 1.7

TEACHING TIP Discuss with your students when a bar graph or a line graph should be used to represent a given set of data. Whereas all line graphs could also be bar graphs, the opposite is not true: some bar graphs do not make sense as line graphs. The idea of change over time needs to be emphasized: if a quantity changes over time it can be graphed either way. Present different sets of data to your students. Ask them what type of graph they would use and why.

LESSON 1.8

COMMON ERROR After reading the examples included in this lesson, some students might jump to the conclusion that the graph of a function will always be a straight line. Graph the data for Example 1 on page 48 to show that functions can have graphs that are not lines.

TEACHING TIP Once students are familiar with the ideas of domain and range, ask students if it makes sense to take any value as an input or output. For instance, in Example 2 would it make sense to take decimal values for the time? What would they mean? Can we expect to get decimal values for the height? This discussion will help students to understand that it is possible to have other values, which will fill in the line between plotted points.

Outside Resources

BOOKS/PERIODICALS

Algebraic Thinking, Grades K–12: Readings from NCTM's School-Based Journals and Other Publications. Helps teachers understand the development of and activities that foster algebraic thinking. Reston, VA; NCTM, 1999.

ACTIVITIES/MANIPULATIVES

Cox, Pam and Linda Bridges. "Algebra for All: Calculating Human Horsepower." *Activities: Mathematics Teacher* (March 1999) pp. 225–228.

SOFTWARE

How the West was One + Three x Four. Order of operations challenge problems. Pleasantville, NY; 1994.

VIDEOS

Mathematics: Making the Connection. How mathematics is used in several professions. Reston, VA; NCTM 1996.

NAME _____ DATE _____

Parent Guide for Student Success

For use with Chapter 1: Connections to Algebra

Chapter Overview One way that you can help your student succeed in Chapter 1 is by discussing the lesson goals in the chart below. When a lesson is completed, ask your student to interpret the lesson goals for you and to explain how the mathematics of the lesson relates to one of the key applications listed in the chart.

Lesson Title	Lesson Goal	Key Applications
1.1: Variables in Algebra	Evaluate variable expressions.	• Race Cars • Driving Distances • Unit Analysis
1.2: Exponents and Powers	Evaluate a power.	• Fish Tanks • Sculpture • Candle Making
1.3: Order of Operations	Use the established order of operations.	• Basketball • Football Uniforms • Admission Prices
1.4: Equations and Inequalities	Check solutions of equations and inequalities.	• Cost of Ingredients • Veterinarian • Computer Center • Aircraft Design
1.5: Translating Words into Mathematical Symbols	Translate words into mathematical symbols.	• Phone Conversations • Land Ordinance of 1785
1.6: A Problem Solving Plan Using Models	Model and solve real-life problems.	• Dining Out • Sports Fields • Jet Pilots
1.7: Tables and Graphs	Organize real-life data using a table or graph.	• Eating Habits • Movie Making • Weather
1.8: An Introduction to Functions	Use four different ways to represent functions.	• Hot-Air Balloon • Apple Trees • Water Temperatures

Study Tip

Keep a Math Notebook is the study tip featured in Chapter 1 (see page 2). Be sure your student has set up a notebook in which to keep math notes. Encourage your student to record notes in math class each day and to review them at home each night. Your student can use these math notes to share with you what was done in class.

Parent Guide for Student Success

For use with Chapter 1

Key Ideas Your student can demonstrate understanding of key concepts by working through the following exercises with you.

Lesson	Exercise
1.1	A car travels 50 miles in 90 minutes. What is the average speed of the car?
1.2	Evaluate $5x^2$ and $(5x)^2$ when $x = 2$.
1.3	The area of a trapezoid can be found with the formula $A = \frac{1}{2}h(b_1 + b_2)$. Find the area of a trapezoid with a height of 9 meters and bases of 7 and 11 meters.
1.4	You are taking a family trip. You need to travel 270 miles and can average 60 miles per hour. Use $60t \geq 270$ to model the situation. What do 60, t, and 270 represent? Can you make the trip in 4 hours? in 5 hours?
1.5	Write the sentence "two more than a number is 8" as an equation or an inequality.
1.6	Write an algebraic model and solve it to find how many juices you bought last week at $.90 each if you spent $5.40.
1.7	In 1996, the average personal income of people living in Kansas was $23,281. In Alabama, it was $20,055; in California, $25,144; in Massachusetts, $29,439; in Montana, $19,047; in New York, $28,782; and in Texas, $22,045. What scale might you use to make a bar graph?
1.8	Find the output values for $x = 0$, 1, and 2 for the function $y = 8 - 3x$.

Home Involvement Activity

Directions: Use exactly four 4's and grouping symbols to make each whole number from one to ten. For example, $1 = \frac{4}{4} + (4 - 4)$.

Activity Sample Answers

$$1 = \frac{4}{4} + (4 - 4) \quad 2 = \frac{4}{4} + \frac{4}{4} \quad 3 = \frac{(4 + 4 + 4)}{4} \quad 4 = 4 + 4 \cdot (4 - 4) \quad 5 = \frac{(4 \cdot 4 + 4)}{4}$$

$$6 = 4 + \left(\frac{4 + 4}{4}\right) \quad 7 = 4 + 4 - \left(\frac{4}{4}\right) \quad 8 = 4 + 4 + (4 - 4) \quad 9 = 4 + 4 + \left(\frac{4}{4}\right) \quad 10 = \frac{(44 - 4)}{4}$$

Answers

1.1: about 33 mi/h **1.2:** 20, 100 **1.3:** 81 m^2 **1.4:** 60 is the speed, t is the driving time, and 270 is the distance traveled; no; yes **1.5:** $x + 2 = 8$ **1.6:** 6 juices **1.7:** *Sample answer: from 0 to $30,000 by $5000* **1.8:** 8, 5, 2

NAME _____ DATE _____

Prerequisite Skills Review

For use before Chapter 1

EXAMPLE 1 *Evaluating Expressions*

Evaluate the expression.

a. $4 \cdot 5 \cdot \dfrac{1}{2}$

b. $\dfrac{1}{8} + \dfrac{3}{8}$

c. $100(0.08)(2.5)$

SOLUTION

a. $4 \cdot 5 \cdot \dfrac{1}{2} = 10$

b. $\dfrac{1}{8} + \dfrac{3}{8} = \dfrac{4}{8} = \dfrac{1}{2}$

c. $100(0.08)(2.5) = 20$

Exercises for Example 1

Evaluate the expression.

1. $12(1.5)(2)$

2. $2.6 + 4.9 + 1.3$

3. $0.05 - 0.02$

4. $\dfrac{1}{3} + \dfrac{2}{3}$

5. $10\left(\dfrac{1}{10}\right)$

6. $\left(\dfrac{1}{2}\right)\left(\dfrac{1}{2}\right)$

EXAMPLE 2 *Working with Inequalities*

Complete the statement using $>$ or $<$.

a. $331 \underline{\ ?\ } 301$

b. $103.01 \underline{\ ?\ } 103.1$

SOLUTION

a. First, plot the numbers on a number line. Because 331 is to the right of 301, it follows that $331 > 301$.

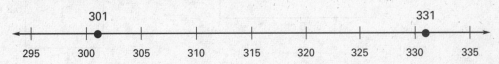

b. First, plot the numbers on a number line. Because 103.01 is to the left of 103.1, it follows that $103.01 < 103.1$.

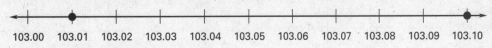

Exercises for Example 2

Complete the statement using > or <.

7. $501 \underline{\ ?\ } 510$

8. $98.02 \underline{\ ?\ } 98.021$

9. $7895 \underline{\ ?\ } 7885$

10. $0.85 \underline{\ ?\ } 0.085$

11. $4041 \underline{\ ?\ } 4401$

12. $0.0022 \underline{\ ?\ } 0.02$

NAME _____ DATE _____

Prerequisite Skills Review

For use before Chapter 1

EXAMPLE 3 *Finding Area and Perimeter*

Find the area and perimeter of the figure.

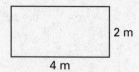

2 m
4 m

SOLUTION

The area can be found as follows.

Area = Length · Width
$$= 4 \cdot 2$$
$$= 8$$

The area is 8 square meters.

The perimeter can be found as follows.

Perimeter = 2 · Length + 2 · Width
$$= 2 \cdot 4 + 2 \cdot 2$$
$$= 8 + 4$$
$$= 12$$

The perimeter is 12 meters.

Exercises for Example 3

Find the area and perimeter of the figure.

13.

2 cm
6 cm

14.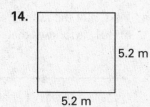

5.2 m
5.2 m

15. 3.1 cm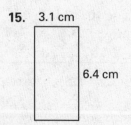

6.4 cm

16. 4.8 cm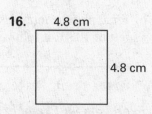

4.8 cm

Algebra 1
Chapter 1 Resource Book

Strategies for Reading Mathematics

For use with Chapter 1

Strategy: Reading an Algebra Textbook

Flip through your algebra book. It doesn't really look like your other textbooks, does it? A history textbook, for example, has long passages to read. An algebra book has short explanations or problems followed by examples and solutions. Here are some tips to help you read your algebra book. Example 1 on page 15 of your text is used to demonstrate.

Look for new vocabulary.

ORDER OF OPERATIONS In arithmetic and algebra there is **order of operations** to evaluate an expression involving more than one operation.

Look for boxed summaries.

> **ORDER OF OPERATIONS**
>
> **STEP ❶** First do operations that occur within grouping symbols.
>
> **STEP ❷** Then evaluate powers.
>
> **STEP ❸** Then do multiplications and divisions from left to right.
>
> **STEP ❹** Finally, do additions and subtractions from left to right.

EXAMPLE 1 *Evaluate Without Grouping Symbols*

Evaluate the expression $3x^2 + 1$ when $x = 4$. Use the order of operations.

Find and read all examples.

SOLUTION

Parts of the solution that are highlighted show key information.

$$3x^2 + 1 = 3 \times 4^2 + 1 \qquad \textbf{Substitute 4 for } x.$$

$$= 3 \times 16 + 1 \qquad \textbf{Evaluate power.} \leftarrow \text{Read labels. Check}$$
$$= 48 + 1 \qquad \qquad \textbf{Multiply 3 times 16.} \quad \text{them against the}$$
$$= 49 \qquad \qquad \quad \textbf{Add.} \qquad \qquad \qquad \text{calculations.}$$

STUDY TIP

Do Calculations

After you have read through a worked-out example, go back and try to recreate the solution on your own. Then check your solution process against the one shown in the book.

STUDY TIP

Read Diagrams

Many lessons include diagrams. Some illustrate new ideas and vocabulary. Others show how to represent a problem. Always read the labels on a diagram. Match the information in the diagram with the written information in the text.

Questions

1. Which mathematical symbols did you need to know to understand the example above?

2. In a worked out example, how can you decide where a given number comes from? For example, in the solution for the example above, where did the 16 come from in Step 2? Where did the 48 come from in Step 3?

3. Use the order of operations to evaluate the expression $3(x^2 + 5) - 1$ when $x = 3$. Show each step in the process you use.

NAME _____ DATE _____

Strategies for Reading Mathematics

For use with Chapter 1

Visual Glossary

The Study Guide on page 2 lists the key vocabulary for Chapter 1. Use the visual glossary below to help you understand some of the key vocabulary in Chapter 1. You may want to copy these diagrams into your notebook and refer to them as you complete the chapter.

> **GLOSSARY**
>
> **variable expression** (p. 3) A collection of constants, variables, and operations.
>
> **power** (p. 9) The result of repeated multiplication. For example, in the expression $4^2 = 16$, 16 is the second power of 4.
>
> **exponent** (p. 9) The number or variable that represents the number of times the base is used as a factor. For example, in the expression 4^6, 6 is the exponent.
>
> **base** (p. 9) The number or variable that is used as a factor in repeated multiplication. For example, in the expression 4^6, 4 is the base.
>
> **equation** (p. 24) A statement formed when an equal sign is placed between two expressions.
>
> **inequality** (p. 26) A statement formed when an inequality symbol is placed between two expressions.

Expressions

Understanding how to represent problems with variable expressions is a fundamental requirement of algebraic thinking.

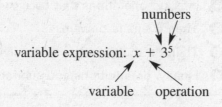

variable expression: $x + 3^5$

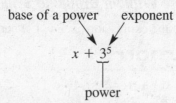

$x + 3^5$

3^5 means 3 to the **fifth** power, or $3 \cdot 3 \cdot 3 \cdot 3 \cdot 3$.

Equations

Many algebraic problems are represented by equations. An equation contains an equal sign ($=$).

equal sign

equations: $5^2 - 1 = 24$
              $2x + 1 = 5$

Inequalities

Some algebraic problems are represented by inequalities. An inequality contains an inequality symbol such as $<$, $\leq$, $>$, or $\geq$.

inequality symbol

inequalities: $7 < 3^2 - 1$
                 $2z - 1 \leq 5$
                 $x > 3$
                 $y^2 \geq 0$

TEACHER'S NAME _____ CLASS _____ ROOM _____ DATE _____

Lesson Plan

1-day lesson (See *Pacing the Chapter,* TE page 1A)   **For use with pages 3–8**

GOAL   Evaluate variable expressions.

State/Local Objectives _____

_____

✓ Check the items you wish to use for this lesson.

STARTING OPTIONS
____ Prerequisite Skills Review: CRB pages 5–6
____ Strategies for Reading Mathematics: CRB pages 7–8
____ Warm-Up: CRB page 11 or Transparencies

TEACHING OPTIONS
____ Lesson Opener: CRB page 12 or Transparencies
____ Examples 1–5: SE pages 3–5
____ Extra Examples: TE pages 4–5 or Transparencies
____ Checkpoint Exercises: SE pages 3–5
____ Graphing Calculator Activity with Keystrokes: CRB pages 13–14
____ Concept Check: TE page 5
____ Guided Practice Exercises: SE page 6

APPLY/HOMEWORK
Homework Assignment
____ Transitional: SRH p. 773 Exs. 1–4; pp. 6–8 Exs. 17–23, 34–37, 40–42, 50, 51, 53–65 odd
____ Average: pp. 6–8 Exs. 24–29, 33–39 odd, 42, 43, 50–62 all
____ Advanced: pp. 6–8 Exs. 27–33, 37–39, 43–47*, 50, 51, 55–57, 64–66; EC: CRB p. 21
Reteaching the Lesson
____ Practice Masters: CRB pages 15–16 (Level A, Level B)
____ Reteaching with Practice: CRB pages 17–18 or Practice Workbook with Examples; Resources in Spanish
____ Personal Student Tutor: CD-ROM
Extending the Lesson
____ Interdisciplinary/Real-Life Applications: CRB page 20
____ Challenge: CRB page 21

ASSESSMENT OPTIONS
____ Daily Quiz (1.1): TE page 8, CRB page 24, or Transparencies
____ Standardized Test Practice: SE page 8; STP Workbook; Transparencies

Notes _____

_____

_____

TEACHER'S NAME _____ CLASS _____ ROOM _____ DATE _____

Lesson Plan for Block Scheduling

Half-block lesson (See *Pacing the Chapter,* TE page 1A)   **For use with pages 3–8**

GOAL   Evaluate variable expressions.

State/Local Objectives _____

_____

_____

CHAPTER PACING GUIDE	
Day	**Lesson**
1	**1.1 (all)**; 1.2 (all)
2	1.3 (all); 1.4 (all)
3	1.5 (all); 1.6 (all)
4	1.7 (all); 1.8 (all)
5	Ch. 1 Review and Assess

✓ **Check the items you wish to use for this lesson.**

STARTING OPTIONS

____ Prerequisite Skills Review: CRB pages 5–6

____ Strategies for Reading Mathematics: CRB pages 7–8

____ Warm-Up: CRB page 11 or Transparencies

TEACHING OPTIONS

____ Lesson Opener: CRB page 12 or Transparencies

____ Examples 1–5: SE pages 3–5

____ Extra Examples: TE pages 4–5 or Transparencies

____ Checkpoint Exercises: SE pages 3–5

____ Graphing Calculator Activity with Keystrokes: CRB pages 13–14

____ Concept Check: TE page 5

____ Guided Practice Exercises: SE page 6

APPLY/HOMEWORK

Homework Assignment (See also the assignment for Lesson 1.2.)

____ Block Schedule: pp. 6–8 Exs. 24–29, 33–39 odd, 42, 43, 50–62 all

Reteaching the Lesson

____ Practice Masters: CRB pages 15–16 (Level A, Level B)

____ Reteaching with Practice: CRB pages 17–18 or Practice Workbook with Examples; Resources in Spanish

____ Personal Student Tutor: CD-ROM

Extending the Lesson

____ Interdisciplinary/Real-Life Applications: CRB page 20

____ Challenge: CRB page 21

ASSESSMENT OPTIONS

____ Daily Quiz (1.1): TE page 8, CRB page 24, or Transparencies

____ Standardized Test Practice: SE page 8; STP Workbook; Transparencies

Notes _____

_____

_____

NAME _____ DATE _____

WARM-UP EXERCISES

For use before Lesson 1.1, pages 3–8

Lesson 1.1

Evaluate the expression.

1. $120(0.05)(3.5)$

2. $\dfrac{3}{4} \cdot \dfrac{1}{3}$

3. $\dfrac{1}{3} + \dfrac{1}{3}$

4. $\dfrac{100}{0.25}$

5. Find the perimeter of the square.

7.1 cm

Lesson 1.1

SET UP: Work in a group. Be sure at least one classmate in your group is not the same age as you.

Here is a fun puzzle for you to solve. When you are finished, you will try to explain why it works!

PUZZLE: Start with your current age.

STEP ❶ Add 5 to the age.

STEP ❷ Multiply the result of Step 1 by 2.

STEP ❸ Subtract twice the age you started with from the result of Step 2.

STEP ❹ Subtract 10 from the result of Step 3.

---

1. Compare your answer with the answers of other members of your group. What do you notice?

2. Choose an age of an adult. Be sure the age you choose is different from the ages chosen by other members of your group. Repeat the puzzle and Question 1 for this age.

3. Choose an age of someone younger than you. Be sure the age you choose is different from the ages chosen by other members of your group. Repeat the puzzle and Question 1 for this age.

4. What conclusion can you make about this puzzle? Explain why you think the puzzle works this way.

5. Work with your group to make up a puzzle of your own. Use different numbers to check your puzzle. Then explain why your puzzle works the way it does.

NAME _____ DATE _____

Graphing Calculator Activity

For use with pages 3–8

GOAL **To use the store feature of the graphing calculator to evaluate variable expressions.**

Evaluating an expression means to replace each variable in the expression with a number and then find the result. Often you are asked to evaluate several expressions for the same number. The store feature of the graphing calculator not only can save you time, but also can help you check your work. This feature is especially useful when the variable expressions become more complex.

Activity

❶ Store the value 3 in the variable X on your calculator.

❷ Evaluate each variable expression by entering it into your calculator.

    a. $8 + x$ **b.** $19x$ **c.** $42 - x$ **d.** $\dfrac{45}{x}$

❸ Store the value 9 in the variable Y on your calculator.

❹ Evaluate each variable expression by entering it into your calculator.

    a. $x + y$ **b.** $23 - x - y$ **c.** $4xy$ **d.** $\dfrac{y}{x}$

Exercises

Evaluate the variable expression when $x = 4$.

1. $x - 1$ **2.** $13x$ **3.** $\dfrac{48}{x}$ **4.** $21 + x$

Evaluate the variable expression when $x = 8$ and $y = \dfrac{1}{2}$.

5. xy **6.** $x - y$ **7.** $4 + y + x$ **8.** $\dfrac{x}{y}$

See page 14 for keystrokes.

Graphing Calculator Activity

For use with pages 3–8

Lesson 1.1

TI-82

3 [STO▸] [X,T,θ] [ENTER]

8 [+] [X,T,θ] [ENTER]

19 [X,T,θ] [ENTER]

42 [−] [X,T,θ] [ENTER]

45 [÷] [X,T,θ] [ENTER]

9 [STO▸] [ALPHA] [Y] [ENTER]

[X,T,θ] [+] [ALPHA] [Y] [ENTER]

23 [−] [X,T,θ] [−] [ALPHA] [Y] [ENTER]

4 [X,T,θ] [ALPHA] [Y] [ENTER]

[ALPHA] [Y] [÷] [X,T,θ] [ENTER]

TI-83

3 [STO▸] [X,T,θ,n] [ENTER]

8 [+] [X,T,θ,n] [ENTER]

19 [X,T,θ,n] [ENTER]

42 [−] [X,T,θ,n] [ENTER]

45 [÷] [X,T,θ,n] [ENTER]

9 [STO▸] [ALPHA] [Y] [ENTER]

[X,T,θ,n] [+] [ALPHA] [Y] [ENTER]

23 [−] [X,T,θ,n] [−] [ALPHA] [Y] [ENTER]

4 [X,T,θ,n] [ALPHA] [Y] [ENTER]

[ALPHA] [Y] [÷] [X,T,θ,n] [ENTER]

SHARP EL-9600c

3 [STO] [X/θ/T/n] [ENTER]

8 [+] [X/θ/T/n] [ENTER]

19 [X/θ/T/n] [ENTER]

42 [−] [X/θ/T/n] [ENTER]

45 [÷] [X/θ/T/n] [ENTER]

9 [STO] [ALPHA] [Y] [ENTER]

[X/θ/T/n] [+] [ALPHA] [Y] [ENTER]

23 [−] [X/θ/T/n] [−] [ALPHA] [Y] [ENTER]

4 [X/θ/T/n] [ALPHA] [Y] [ENTER]

[ALPHA] [Y] [÷] [X/θ/T/n] [ENTER]

CASIO CFX-9850GA PLUS

From the main menu, choose RUN.

3 [→] [X,θ,T] [EXE]

8 [+] [X,θ,T] [EXE]

19 [X,θ,T] [EXE]

42 [−] [X,θ,T] [EXE]

45 [÷] [X,θ,T] [EXE]

9 [→] [ALPHA] [Y] [EXE]

[X,θ,T] [+] [ALPHA] [Y] [EXE]

23 [−] [X,θ,T] [−] [ALPHA] [Y] [EXE]

4 [X,θ,T] [ALPHA] [Y] [EXE]

[ALPHA] [Y] [÷] [X,θ,T] [EXE]

NAME _____ DATE _____

Practice A

For use with pages 3–8

Name the operation indicated by the expression.

1. $7 - m$ **2.** $12 \div x$ **3.** $t(12)$ **4.** $\dfrac{m}{3}$

Evaluate the expression for the given value of the variable.

5. $x + 5$ when $x = 4$ **6.** $y - 4$ when $y = 19$ **7.** $3x$ when $x = 25$

8. $w + 8$ when $w = 36$ **9.** $7 - x$ when $x = 2$ **10.** $24 \div a$ when $a = 6$

11. $\dfrac{g}{8}$ when $g = 72$ **12.** $\dfrac{m}{3}$ when $m = 27$ **13.** $\dfrac{5}{8} \cdot t$ when $t = 16$

14. $h(25)$ when $h = 100$ **15.** $\dfrac{2}{3} + x$ when $x = \dfrac{2}{3}$ **16.** $\dfrac{24}{y}$ when $y = 6$

Find the distance traveled using *d = rt*.

17. An airplane travels at a rate of 6 miles per minute for 120 minutes.

18. A friend jogs at a rate of 4 miles per hour for 30 minutes.

19. A car travels at a rate of 96 kilometers per hour for 3 hours.

In Exercises 20 and 21, use the diagram below.

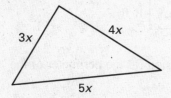

20. Write an expression for the perimeter of the triangle.

21. Find the perimeter if $x = 4$ inches.

In Exercises 22 and 23, use the diagram below.

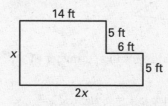

22. Write an expression for the perimeter of the figure shown.

23. Find the perimeter if $x = 10$ feet.

Algebra 1
Chapter 1 Resource Book

NAME _____ DATE _____

Practice B

For use with pages 3–8

Evaluate the expression for the given value of the variable.

1. $13x$ when $x = 4$

2. $y - 4$ when $y = 23$

3. $5x$ when $x = 4$

4. $w + 8$ when $w = 36$

5. $x(7)$ when $x = 29$

6. $24 \div a$ when $a = 6$

7. $\dfrac{g}{12}$ when $g = 72$

8. $\dfrac{78}{m}$ when $m = 26$

9. $8 \cdot t$ when $t = 4$

10. $h(25)$ when $h = 1$

11. $\dfrac{2}{3} + x$ when $x = 2\dfrac{2}{3}$

12. $\dfrac{24}{y}$ when $y = 8$

13. $(34)x$ when $x = 0$

14. $201 + y$ when $y = 39$

15. $8w$ when $w = 7$

Find the distance traveled using $d = rt$.

16. An airplane travels at a rate of 7 miles per minute for 120 minutes.

17. A friend jogs at a rate of 5 miles per hour for 30 minutes.

18. A car travels at a rate of 98 kilometers per hour for 2 hours.

In Exercises 19 and 20, use the diagram below.

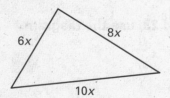

19. Write an expression for the perimeter of the triangle.

20. Find the perimeter if $x = 8$ inches.

In Exercises 21 and 22, use the diagram below.

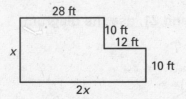

21. Write an expression for the perimeter of the figure shown.

22. Find the perimeter if $x = 20$ feet.

23. *Burning Calories* A 140-pound person playing tennis burns 5 calories per minute. If the person plays for 30 minutes, how many calories does the person burn?

24. *Burning Calories* A 140-pound person in-line skating burns 6 calories per minute. If the person skates for 50 minutes how many calories does the person burn?

NAME _____ DATE _____

Reteaching with Practice

For use with pages 3–8

GOAL **Evaluate variable expressions.**

> ### VOCABULARY
>
> A **variable** is a letter that is used to represent a range of numbers.
>
> A **variable expression** consists of constants, variables, and operations.
>
> To **evaluate** a variable expression, you write the expression, substitute the given number for each variable, and simplify.

EXAMPLE 1 *Describe a Variable Expression*

VARIABLE EXPRESSION	MEANING	OPERATION
$3x, 3 \cdot x, (3)(x)$	3 times x	Multiplication
$\dfrac{3}{x}, 3 \div x$	3 divided by x	Division
$3 + x$	3 plus x	Addition
$3 - x$	3 minus x	Subtraction

Exercises for Example 1

State the meaning of the variable expression and name the operation.

1. $4 - x$　　　　**2.** $7y$　　　　**3.** $x + 1$　　　　**4.** $\dfrac{y}{2}$

EXAMPLE 2 *Evaluate a Variable Expression*

Evaluate the expression when $y = 3$.

a. $y - 1$　　　　　　　　　　　　**b.** $12y$

SOLUTION

a. $y - 1 = 3 - 1$　　Substitute 3 for y.　　**b.** $12y = 12(3)$　　Substitute 3 for y.

　　　$= 2$　　　　　　Simplify.　　　　　　　　$= 36$　　　　Simplify.

Exercises for Example 2

Evaluate the expression for the given value of the variable.

5. $8 + x$ when $x = 6$　　**6.** $\dfrac{10}{s}$ when $s = 2$　　**7.** $2a$ when $a = 20$

8. $9 - y$ when $y = 1$　　**9.** $9q$ when $q = 12$　　**10.** $8b$ when $b = 3$

EXAMPLE 3 *Evaluate d = rt to Find Distance*

Find the distance d traveled if you drive at an average speed of 65 miles per hour for 5 hours.

NAME _____ DATE _____

Reteaching with Practice

For use with pages 3–8

SOLUTION

$$d = rt \qquad \text{Write formula}$$

$$= 65(5) \qquad \text{Substitute 65 for } r \text{ and 5 for } t.$$

$$= 325 \qquad \text{Simplify.}$$

The distance traveled is 325 miles.

Exercises for Example 3

11. Find the distance traveled if you drive at an average speed of 65 miles per hour for 4 hours.

12. Find the distance traveled by an airplane flying at an average speed of 350 miles per hour for 6 hours.

EXAMPLE 4   *Finding Area and Perimeter*

a. The area A of a triangle is equal to half the base b times the height h:

$A = \frac{1}{2}bh$. Find the area of the triangle below in square feet.

b. The perimeter P of a triangle is equal to the sum of the lengths of its sides:

$P = a + b + c$. Find the perimeter of the triangle below in feet.

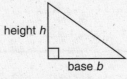

SOLUTION

a. $A = \frac{1}{2}bh$ Write formula.

$A = \frac{1}{2}(4)(3)$ Substitute.

$A = 6$ Simplify.

The triangle has an area of 6 square feet.

b. $P = a + b + c$ Write formula.

$P = 3 + 4 + 5$ Substitute.

$P = 12$ Simplify.

The triangle has a perimeter of 12 feet.

Exercises for Example 4

13. Find the area of a triangle with a base of 10 inches and a height of 6 inches.

14. Find the perimeter of a triangle with side lengths 3 inches, 9 inches, and 11 inches.

NAME _____ DATE _____

Quick Catch-Up for Absent Students

For use with pages 3–8

The items checked below were covered in class on (date missed) _____

Lesson 1.1: Variables in Algebra

____ **Goal:** Evaluate variable expressions.

Material Covered:

_____ Example 1: Describe the Variable Expression

_____ Student Help: Writing Algebra

_____ Example 2: Evaluate the Variable Expression

_____ Example 3: Evaluate *rt* to Find Distance

_____ Student Help: Vocabulary Tip

_____ Example 4: Find the Perimeter

_____ Example 5: Estimate the Area

Vocabulary:

variable, p. 3 ·   values, p. 3

variable expression, p. 3   numerical expression, p. 3

_____ Other (specify) _____

_____

Homework and Additional Learning Support

_____ Textbook (specify) pp. 6–8 _____

_____

_____ *Reteaching with Practice* worksheet (specify exercises)_____

_____ *Personal Student Tutor* for Lesson 1.1

NAME _____ DATE _____

Real-Life Application: When Will I Ever Use This?

For use with pages 3–8

Class Officer Duties:

As class officers, one of the duties for you and your friends is to make sure your class has enough money for your five-year class reunion. You spend the year coming up with and holding various fund raisers. By the end of the year you have earned $1250 in profit. A local bank offers you an incredibly good interest rate of 0.085 in a certificate of deposit.

1. Using the simple interest formula, $I = Prt,$ find the amount of interest that you will have earned by your five-year reunion. In the formula, I stands for interest, r stands for rate, and t stands for time in years. (*Hint*: Assume you have three years left of high school to earn interest, too.) Be sure to show all work when evaluating the expression: write the expression, substitute the known values, and simplify.

2. What is the total amount of money that will be available for you and your classmates when you plan your five-year reunion? (Explain in words how you came up with your answer.)

3. When planning your five-year class reunion, you discover that the hall rental, catering, and door prizes will cost $5000. You decide to charge each person $12.50 to attend the reunion. If you do not use any of the money from your certificate of deposit, how many people p will have to attend if you want all expenses to be covered? Evaluate the expression

$$p = \frac{T}{c},$$

where c is the cost you will charge each person and T is the total amount of money that you need.

4. If you use the money from your certificate of deposit to help pay for the expenses, how many people will need to attend? (*Hint*: You will need to change your total amount in the formula you used from Exercise 3.)

NAME _____ DATE _____

Challenge: Skills and Applications

For use with pages 3–8

1. One mile is approximately equal to 1.609 kilometers. Use unit analysis to change 60 miles per hour into kilometers per hour (km/h).

2. Use unit analysis to change 60 miles per hour into feet per second.

3. The speed of light is about 186,282 miles per second. Earth is about 92,976,000 miles from the sun. How long does it take the sun's light to reach Earth, to the nearest hundredth of a minute?

4. When interest is compounded quarterly, the interest at the end of the quarter is added to the principal before interest is computed for the second quarter. Complete the table below to find the balance after 2 years for $250 at an annual rate of 0.035, or 3.5%, compounded quarterly. Round to the nearest cent whenever necessary. The formula for interest is $I = Prt$.

Period	Beginning balance	Interest	Ending balance
1	$250	$2.19	$252.19
2	$252.19	$2.21	$254.40
3	$254.40		
4			
5			
6			
7			
8			

5. How much simple interest would you earn on $250 at an annual rate of 0.035 for 2 years? How does this compare with the amount of compound interest from Exercise 4?

TEACHER'S NAME _____ CLASS _____ ROOM _____ DATE _____

Lesson Plan

1-day lesson (See *Pacing the Chapter,* TE page 1A)   **For use with pages 9–14**

GOAL **Evaluate a power.**

State/Local Objectives _____

_____

✓ Check the items you wish to use for this lesson.

STARTING OPTIONS

_____ Homework Check (1.1): TE page 6; Answer Transparencies

_____ Homework Quiz (1.1): TE page 8, CRB page 24, or Transparencies

_____ Warm-Up: CRB page 24 or Transparencies

TEACHING OPTIONS

_____ Lesson Opener: CRB page 25 or Transparencies

_____ Examples 1–5: SE pages 9–11

_____ Extra Examples: TE pages 10–11 or Transparencies; Internet Help at *www.mcdougallittell.com*

_____ Checkpoint Exercises: SE pages 9–11

_____ Concept Check: TE page 11

_____ Guided Practice Exercises: SE page 12

APPLY/HOMEWORK

Homework Assignment

_____ Transitional: pp. 12–14 Exs. 13–15, 20–23, 28–30, 34–36, 40–42, 52, 53, 61–63, 65–93 odd

_____ Average: pp. 12–14 Exs. 16–18, 24–27, 37–42, 46–48, 54–57, 61–63, 64–69, 76–83, 88–90

_____ Advanced: pp. 12–14 Exs. 19, 31–33, 37–39, 43–45, 49–51, 54–63, 73–75, 84–87, 91–93;
EC: CRB p. 32

Reteaching the Lesson

_____ Practice Masters: CRB pages 26–27 (Level A, Level B)

_____ Reteaching with Practice: CRB pages 28–29 or Practice Workbook with Examples;
Resources in Spanish

_____ Personal Student Tutor: CD-ROM

Extending the Lesson

_____ Interdisciplinary/Real-Life Applications: CRB page 31

_____ Challenge: CRB page 32

ASSESSMENT OPTIONS

_____ Daily Quiz (1.2): TE page 14, CRB page 35, or Transparencies

_____ Standardized Test Practice: SE page 14; STP Workbook; Transparencies

Notes _____

_____

_____

TEACHER'S NAME _____ CLASS _____ ROOM _____ DATE _____

Lesson Plan for Block Scheduling

Half-block lesson (See *Pacing the Chapter*, TE page 1A)   For use with pages 9–14

GOAL **Evaluate a power.**

State/Local Objectives _____

_____

_____

✓ **Check the items you wish to use for this lesson.**

STARTING OPTIONS

____ Homework Check (1.1): TE page 6; Answer Transparencies

____ Homework Quiz (1.1): TE page 8, CRB page 24, or Transparencies

____ Warm-Up: CRB page 24 or Transparencies

TEACHING OPTIONS

____ Lesson Opener: CRB page 25 or Transparencies

____ Examples 1–5: SE pages 9–11

____ Extra Examples: TE pages 10–11 or Transparencies; Internet Help at www.mcdougallittell.com

____ Checkpoint Exercises: SE pages 9–11

____ Concept Check: TE page 11

____ Guided Practice Exercises: SE page 12

APPLY/HOMEWORK

Homework Assignment (See also the assignment for Lesson 1.1.)

____ Block Schedule: pp. 12–14 Exs. 16–18, 24–27, 37–42, 46–48, 54–57, 61–63, 64–69, 76–83, 88–90

Reteaching the Lesson

____ Practice Masters: CRB pages 26–27 (Level A, Level B)

____ Reteaching with Practice: CRB pages 28–29 or Practice Workbook with Examples; Resources in Spanish

____ Personal Student Tutor: CD-ROM

Extending the Lesson

____ Interdisciplinary/Real-Life Applications: CRB page 31

____ Challenge: CRB page 32

ASSESSMENT OPTIONS

____ Daily Quiz (1.2): TE page 14, CRB page 35, or Transparencies

____ Standardized Test Practice: SE page 14; STP Workbook; Transparencies

Notes _____

_____

_____

CHAPTER PACING GUIDE	
Day	**Lesson**
1	1.1 (all); **1.2 (all)**
2	1.3 (all); 1.4 (all)
3	1.5 (all); 1.6 (all)
4	1.7 (all); 1.8 (all)
5	Ch. 1 Review and Assess

NAME _____ DATE _____

WARM-UP EXERCISES

For use before Lesson 1.2, pages 9–14

Evaluate the expression.

1. $2 \cdot 2 \cdot 2 \cdot 2$

2. $3x$ when $x = 6$

3. $(x)(x)(x)$ when $x = 4$

4. $x - 8$ when $x = 17$

5. $y \cdot y \cdot y \cdot y \cdot y$ when $y = 3$

DAILY HOMEWORK QUIZ

For use after Lesson 1.1, pages 3–8

State the meaning of the variable expression.

1. $15 + x$

2. $\dfrac{45}{n}$

Evaluate the variable expression when $a = 12$.

3. $2a$

4. $\dfrac{a}{4}$

5. A jogger runs 0.75 hour at 8 miles per hour. Find the distance traveled using $d = rt$.

NAME _____ DATE _____

Calculator Lesson Opener

For use with pages 9–14

In the tables below, various numbers are raised to different powers. In Exercise 1, the number 2 is raised to the powers 1, 2, 3, 4, 5, 6, and 7. In the expression 2^3, the 3 is the exponent. You can use a calculator to evaluate expressions like 2^3. If your calculator has a $\boxed{y^x}$ key, enter 2 $\boxed{y^x}$ 3. If your calculator has a $\boxed{\wedge}$ key, enter 2 $\boxed{\wedge}$ 3.

Use a calculator to evaluate the expression in the top row of the table. Write your answer in the bottom row.

1.

2^1	2^2	2^3	2^4	2^5	2^6	2^7

2.

3^1	3^2	3^3	3^4	3^5	3^6	3^7

3.

4^1	4^2	4^3	4^4	4^5	4^6	4^7

4.

5^1	5^2	5^3	5^4	5^5	5^6	5^7

5. Use the tables you have completed to make a conjecture about the meaning of exponents.

NAME _____ DATE _____

Practice A

For use with pages 9–14

Write the expression in exponential form.

1. three to the fourth power   2. seven squared   3. *x* cubed

4. *y* to the fifth power   5. *w* to the third power   6. *a* cubed

7. $4 \cdot 4 \cdot 4 \cdot 4 \cdot 4$ 8. $a \cdot a$ 9. $2 \cdot 2 \cdot 2$

10. $x \cdot x \cdot x \cdot x \cdot x \cdot x$ 11. $5 \cdot 5 \cdot 5 \cdot 5$ 12. $3x \cdot 3x$

Evaluate the power.

13. 4^2 14. 1^6 15. 9^2

16. 2^4 17. 3^3 18. 10^5

Evaluate the expression for the given value(s) of the variable(s).

19. x^3 when $x = 2$ 20. a^2 when $a = 10$

21. x^2 when $x = 5$ 22. $14 - y^2$ when $y = 3$

23. $(x - y)^4$ when $x = 5$ and $y = 3$ 24. $a^2 + b^3$ when $a = 7$ and $b = 1$

25. **Floor Area** Jeff plans to cover the floor of his room with 1 foot square tiles. If the room is a square that measures 10 feet on each side, how many tiles will he need?

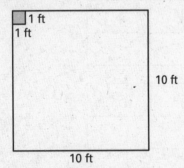

26. **Cubical Package** A cubical box is constructed in order to package a gift. If an edge of the cube is 8 inches, how much material is needed to make the box? (The surface area of a cube is $S = 6s^2$ where s is the edge length.)

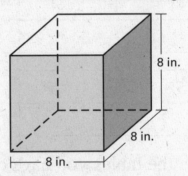

27. **Safe Storage** A safe has a cubical storage space inside. What is the volume of a safe with an interior length of 2 feet?

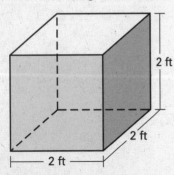

28. **Area Rug** A circular area rug has a radius of 3 feet. How much area does the rug cover? (The area of a circle is $A = \pi r^2$ where $\pi \approx 3.14$ and r is the radius.)

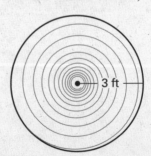

NAME _____ DATE _____

Practice B
For use with pages 9–14

Write the expression in exponential form.

1. three to the fifth power   **2.** nine squared   **3.** y to the seventh power

4. a to the fourth power   **5.** g cubed   **6.** $4 \cdot 4 \cdot 4 \cdot 4 \cdot 4 \cdot b \cdot b$

7. $a \cdot a \cdot a$ **8.** $x \cdot x \cdot x \cdot x \cdot x \cdot x \cdot x \cdot x$ **9.** $3x \cdot 3x \cdot 3x \cdot 3x$

Evaluate the power.

10. 6^2 **11.** 5^3 **12.** 12^2

13. 4^4 **14.** 3^4 **15.** 10^6

Evaluate the expression for the given value(s) of the variable(s).

16. x^3 when $x = 6$ **17.** a^3 when $a = 10$

18. y^2 when $y = 5$ **19.** $6x^2$ when $x = 5$

20. $24 - y^2$ when $y = 3$ **21.** $(x - y)^4$ when $x = 10$ and $y = 3$

22. $a^2 + b^3$ when $a = 7$ and $b = 8$ **23.** $a + b^3$ when $a = 3$ and $b = 2$

24. $(7x - 8)^2$ when $x = 2$ **25.** $(2y)^2 - x^2$ when $x = 3$ and $y = 2$

26. *Safe Storage*   A safe has a cubical storage space inside. What is the volume of a safe with an interior length of 14 inches?

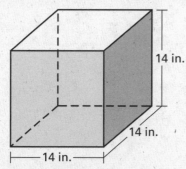

14 in.
14 in.
14 in.

27. *Area Rug*   A circular area rug has a radius of 2 feet. How much area does the rug cover? (The area of a circle is $A = \pi r^2$ where $\pi \approx 3.14$ and r is the radius.)

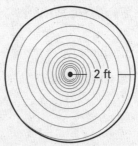

2 ft

28. *Beach Ball*   When blown up, a beach ball has a radius of 12 inches. How much air is needed to blow up the beach ball? (The volume of a sphere is $V = \frac{4}{3}\pi r^3$ where $\pi \approx 3.14$ and r is the radius.)

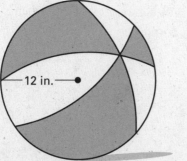

12 in.

29. *Cylindrical Can*   A can has a radius of 2 inches and a height of 7 inches. What is the volume of the can? (The volume of a cylinder is $V = \pi r^2 h$ where $\pi \approx 3.14$, r is the radius, and h is the height.)

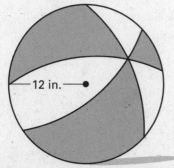

2 in.
7 in.

Lesson 1.2

NAME _____ DATE _____

Reteaching with Practice

For use with pages 9–14

GOAL Evaluate a power.

VOCABULARY

An expression like 2^3 is called a **power,** where the **exponent** 3 represents the number of times the **base** 2 is used as a factor.

Grouping symbols, such as parentheses or brackets, indicate the order in which operations should be performed.

EXAMPLE 1 *Read and Write Powers*

Express the power in words. Then write the meaning.

Exponential Form	Word	Meaning
a. 2^4	two to the fourth power	$2 \cdot 2 \cdot 2 \cdot 2$
b. x^2	x squared	$x \cdot x$

Exercises for Example 1

Express the power in words. Then write the meaning.

1. 3^3 **2.** 5^2 **3.** x^3

EXAMPLE 2 *Evaluate Powers*

Evaluate the expression y^4 when $y = 3$.

SOLUTION

$y^4 = 3^4$ Substitute 3 for y.

 $= 3 \cdot 3 \cdot 3 \cdot 3$ Write factors.

 $= 81$ Multiply.

The value of the expression is 81.

Exercises for Example 2

Evaluate the expression for the given value of the variable.

4. q^3 when $q = 10$ **5.** b^5 when $b = 2$ **6.** z^2 when $z = 5$

7. x^4 when $x = 6$ **8.** m^3 when $m = 9$ **9.** n^5 when $n = 3$

NAME _____ DATE _____

Reteaching with Practice

For use with pages 9–14

EXAMPLE 3 **Exponents and Grouping Symbols**

Evaluate the expression when $x = 2$.

a. $3x^4$ **b.** $(3x)^4$

SOLUTION

a. $3x^4 = 3(2^4)$ Substitute 2 for x.   **b.** $(3x)^4 = (3 \cdot 2)^4$ Substitute 2 for x.

         $= 3(16)$ Evaluate power. $= 6^4$ Multiply within Parentheses.

         $= 48$ Multiply. $= 1296$ Evaluate power.

Exercises for Example 3

Evaluate the expression for the given values of the variables.

10. $(c + d)^3$ when $c = 2$ and $d = 5$ **11.** $c^3 + d^3$ when $c = 2$ and $d = 5$

12. $5p^2$ when $p = 2$ **13.** $(5p)^2$ when $p = 2$

EXAMPLE 4

Find Volume

A storage crate has the shape of a cube. Each edge of the crate is 5 feet long. Find the volume of the crate in cubic feet.

SOLUTION

$V = s^3$ Write formula for volume.

   $= 5^3$ Substitute 5 for s.

   $= 125$ Evaluate power.

The volume of the storage crate is 125 ft^3.

Exercises for Example 4

14. The formula for the area of a square is $A = s^2$. Find the area of a square when $s = 10$ ft.

15. The formula for the area of a square is $A = s^2$. Find the area of a square when $s = 12$ cm.

Lesson 1.2

NAME _____ DATE _____

Quick Catch-Up for Absent Students

For use with pages 9–14

The items checked below were covered in class on (date missed) _____

Lesson 1.2: Exponents and Powers

_____ **Goal:** Evaluate a power.

Material Covered:

_____ Example 1: Read and Write Powers

_____ Example 2: Evaluate the Power

_____ Example 3: Evaluate Exponential Expressions

_____ Example 4: Exponents and Grouping Symbols

_____ Example 5: Find the Volume of the Tank

Vocabulary:

power, p. 9   exponent, p. 9

base, p. 9   grouping symbols, p. 10

Homework and Additional Learning Support

_____ Textbook (specify) <u>pp. 12–14</u> _____

_____

_____ Internet: Extra Examples at www.mcdougallittell.com

_____ *Reteaching with Practice* worksheet (specify exercises) _____

_____ *Personal Student Tutor* for Lesson 1.2

NAME _____ DATE _____

Interdisciplinary Application

For use with pages 9–14

Natatorium

GEOMETRY Competitive pools fall into two categories: long-course pools and short-course pools. Long-course pools (also known as Olympic-sized pools) are 50 meters long and 25 meters wide. In national and international competitions, the water must be between 78 and 80 degrees Fahrenheit and a minimum of four feet deep. Pools used in the Olympics must be 2.0 meters deep. Short-course pools are 22.9 meters long and 16 meters wide. These pools are used in most high school and college competitions. Each pool is separated into lanes marked by plastic markers called lane-lines. These markers not only separate the swimmers from each other, but also help to keep the surface of the water calm by absorbing the waves of the swimmers. A touch pad at the end of each lane records a swimmer's performance to the hundredth of a second.

1. Use the above information to find the volume in cubic meters of a pool that could be used for the Olympics. Find the volume using the formula for the volume of a rectangular prism. The formula is the length times the width times the height.

2. If there are 3.28 feet in 1 meter, how many cubic feet are in the pool in Exercise 1?

3. Converting to liquid volume, where one cubic foot holds 7.48 gallons, how many gallons of water can the Olympic pool hold (from Exercise 1)?

4. How much area would an Olympic-sized pool cover?

5. How much area would a short-course pool cover?

6. A pool used to warm up and cool down after swimming has the shape of a cube. If each side of the pool is 6.5 feet, what is the volume?

7. If one cubic foot holds 7.48 gallons of water, how many gallons will the cubic pool from Exercise 6 hold?

NAME _____ DATE _____

Challenge: Skills and Applications

For use with pages 9–14

Evaluate each expression for the given values of the variables.

1. a^n when $a = 4$ and $n = 3$

2. $(a - 5)^n$ when $a = 7$ and $n = 5$

3. $(9 - a)^n$ when $a = 6$ and $n = 4$

Find the value of *n* in each.

4. $2^n = 16$ **5.** $3^n = 9$ **6.** $10^n = 10,000,000$ **7.** $5^n = 625$

8. If $3^{-2} = \frac{1}{9}$, what do you think 4^{-2} equals?

9. If $\left(\frac{1}{2}\right)^{-3} = 8$, what do you think $\left(\frac{1}{2}\right)^{-4}$ equals?

For Exercises 10–12, use the following figures.

Figure 1   Figure 2   Figure 3

10. Copy the first three figures on graph paper. Then draw the fourth and the fifth figures of the sequence.

11. Look for a pattern in the areas of the figures and complete the table.

Figure	1	2	3	4	5
Area	5	8	13		
Pattern					

12. Write an expression using an exponent for the area of the *n*th figure.

Algebra 1
Chapter 1 Resource Book

TEACHER'S NAME _____ CLASS _____ ROOM _____ DATE _____

Lesson Plan

1-day lesson (See *Pacing the Chapter*, TE pages 1A)   **For use with pages 15–21**

GOAL **Use the established order of operations.**

State/Local Objectives _____

_____

✓ **Check the items you wish to use for this lesson.**

STARTING OPTIONS
____ Homework Check (1.2): TE page 12; Answer Transparencies
____ Homework Quiz (1.2): TE page 14, CRB page 35, or Transparencies
____ Warm-Up: CRB page 35 or Transparencies

TEACHING OPTIONS
____ Lesson Opener: CRB page 36 or Transparencies
____ Examples 1–5: SE pages 15–17
____ Extra Examples: TE pages 16–17 or Transparencies
____ Checkpoint Exercises: SE pages 15–17
____ Graphing Calculator Activity with Keystrokes: CRB pages 37–38
____ Concept Check: TE page 17
____ Guided Practice Exercises: SE page 18

APPLY/HOMEWORK
Homework Assignment
____ Transitional: SRH p. 766 Exs. 49–56; pp. 18–21 Exs. 17–19, 23, 24, 27–29, 37–41 odd, 42–44, 54–60, 64–73, Quiz 1
____ Average: pp. 18–21 Exs. 20–22, 25, 26, 30–32, 36–38, 42, 45, 46, 51, 52, 54–57, 59–77 odd, Quiz 1
____ Advanced: pp. 18–21 Exs. 33–35, 39–42, 47–59*, 67–69, 74–77, Quiz 1; EC: CRB p. 45

Reteaching the Lesson
____ Practice Masters: CRB pages 39–40 (Level A, Level B)
____ Reteaching with Practice: CRB pages 41–42 or Practice Workbook with Examples; Resources in Spanish
____ Personal Student Tutor: CD-ROM

Extending the Lesson
____ Interdisciplinary/Real-Life Applications: CRB page 44
____ Challenge: CRB page 45

ASSESSMENT OPTIONS
____ Daily Quiz (1.3): TE page 21, CRB page 49, or Transparencies
____ Standardized Test Practice: SE page 20; STP Workbook; Transparencies
____ Quiz 1.1–1.3: SE page 21; CRB page 46; Resources in Spanish

Notes _____

_____

_____

TEACHER'S NAME _____ CLASS _____ ROOM _____ DATE _____

Lesson Plan for Block Scheduling

Half-block lesson (See *Pacing the Chapter,* TE pages 1A)   For use with pages 15–21

GOAL   Use the established order of operations.

State/Local Objectives _____

_____

_____

✓ **Check the items you wish to use for this lesson.**

STARTING OPTIONS

____ Homework Check (1.2): TE page 12; Answer Transparencies

____ Homework Quiz (1.2): TE page 14,
        CRB page 35, or Transparencies

____ Warm-Up: CRB page 35 or Transparencies

TEACHING OPTIONS

____ Lesson Opener: CRB page 36 or Transparencies

____ Examples 1–5: SE pages 15–17

____ Extra Examples: TE pages 16–17 or Transparencies

____ Checkpoint Exercises: SE pages 15–17

____ Graphing Calculator Activity with Keystrokes: CRB pages 37–38

____ Concept Check: TE page 17

____ Guided Practice Exercises: SE page 18

APPLY/HOMEWORK

Homework Assignment (See also the assignment for Lesson 1.4.)

____ Block Schedule: pp. 18–21 Exs. 20–22, 25, 26, 30–32, 36–38, 42. 45. 46. 51. 52. 54–57,
        59–77 odd, Quiz 1

Reteaching the Lesson

____ Practice Masters: CRB pages 39–40 (Level A, Level B)

____ Reteaching with Practice: CRB pages 41–42 or Practice Workbook with Examples;
        Resources in Spanish

____ Personal Student Tutor: CD-ROM

Extending the Lesson

____ Interdisciplinary/Real-Life Applications: CRB page 44

____ Challenge: CRB page 45

ASSESSMENT OPTIONS

____ Daily Quiz (1.3): TE page 21, CRB page 49, or Transparencies

____ Standardized Test Practice: SE page 20; STP Workbook; Transparencies

____ Quiz 1.1–1.3: SE page 21; CRB page 46; Resources in Spanish

Notes _____

_____

_____

CHAPTER PACING GUIDE	
Day	**Lesson**
1	1.1 (all); 1.2 (all)
2	**1.3 (all)**; 1.4 (all)
3	1.5 (all); 1.6 (all)
4	1.7 (all); 1.8 (all)
5	Ch. 1 Review and Assess

Lesson 1.3

NAME _____ DATE _____

WARM-UP EXERCISES

For use before Lesson 1.3, pages 15–21

Evaluate the expression.

1. $(3 \cdot 5) + 4$

2. $3 \cdot (5 + 4)$

3. $(9 - 3)^2$

4. $9 - (3^2)$

5. $4w^2$ when $w = 5$

6. $(4w)^2$ when $w = 5$

DAILY HOMEWORK QUIZ

For use after Lesson 1.2, pages 9–14

1. Write k squared in exponential form.

Evaluate the variable expression when $x = 3$ and $y = 2$.

2. x^4

3. $(x + y)^3$

4. $(4x)^2$

5. A square garden is 11 meters on a side. Find the area in square meters. Use the formula for the area A of a square with side of length s: $A = s^2$.

Lesson 1.3

NAME _____ DATE _____

Application Lesson Opener

For use with pages 15–21

1. You are renting a bicycle. You must pay a deposit of $5 plus an additional $6 an hour. You rent a bicycle for 4 hours. Which is the correct way to compute your total bill? Why?

   A. $5 + 6 \times 4 = 5 + 24 = 29$

   B. $5 + 6 \times 4 = 11 \times 4 = 44$

2. You are ordering a pair of jeans from a catalog. The catalog company charges sales tax on the total purchase amount, plus a shipping fee. The jeans you are ordering cost $20. The sales tax rate is 0.05 and the shipping fee is $3. Which is the correct way to compute your total bill? Why?

   A. $20 + 20 \times 0.05 + 3 = 40 \times 0.05 + 3 = 2 + 3 = 5$

   B. $20 + 20 \times 0.05 + 3 = 20 + 1 + 3 = 21 + 3 = 24$

   C. $20 + 20 \times 0.05 + 3 = 40 \times 0.05 + 3 = 40 \times 3.05 = 122$

   D. $20 + 20 \times 0.05 + 3 = 20 + 20 \times 3.05 = 20 + 61 = 81$

3. Consider your answers to Questions 1 and 2. Why do you think the order in which you complete the operations is important?

NAME _____ DATE _____

Graphing Calculator Activity

For use with pages 15–21

GOAL **To enter expressions that include grouping symbols into the graphing calculator correctly.**

The order of operations is an important mathematical process. A graphing calculator is able to perform the operations in the proper order, so grouping symbols must be used correctly to change the order of operations in an expression. The operations inside grouping symbols are always performed first.

Activity

1 Evaluate each expression in your graphing calculator. Explain why the answers differ.

a. $10 + 8 \div 2 - 1$ **b.** $(10 + 8) \div 2 - 1$

2 An expression that uses a fraction bar is often evaluated incorrectly. The fraction bar actually acts as a grouping symbol. Use your calculator to determine which one of the choices below results in the same answer as the evaluated expression given. Explain the difference between the two choices.

$$\frac{48 - 12}{2^3 + 4 - 3} = 4$$

a. $48 - 12 \div 2^3 + 4 - 3$ **b.** $(48 - 12) \div (2^3 + 4 - 3)$

3 Evaluate each expression using the correct grouping symbols. They should result in the same number.

a. $15 - (5^2 - 3) \div 2$ **b.** $\dfrac{7^2 + 3}{20 - 3^2 + 2}$

Exercises

Evaluate the expression using your graphing calculator.

1. $16 \div (8 - 4) + 9$

2. $19 - (2^4 + 3)$

3. $[12 + (6^2 - 6)] \div 2$

4. $\dfrac{7 \cdot 8}{2^3 - 1}$

5. $\dfrac{23 - 3}{30 - 3^3 + 7}$

6. $\dfrac{23 + 19}{19 - (4^2 + 2) + 2}$

See page 38 for keystrokes.

NAME _____ DATE _____

Graphing Calculator Activity

For use with pages 15–21

TI-82

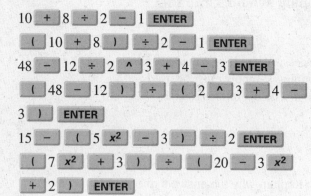

10 `+` 8 `÷` 2 `−` 1 `ENTER`

`(` 10 `+` 8 `)` `÷` 2 `−` 1 `ENTER`

48 `−` 12 `÷` 2 `^` 3 `+` 4 `−` 3 `ENTER`

`(` 48 `−` 12 `)` `÷` `(` 2 `^` 3 `+` 4 `−`

3 `)` `ENTER`

15 `−` `(` 5 `x²` `−` 3 `)` `÷` 2 `ENTER`

`(` 7 `x²` `+` 3 `)` `÷` `(` 20 `−` 3 `x²`

`+` 2 `)` `ENTER`

TI-83

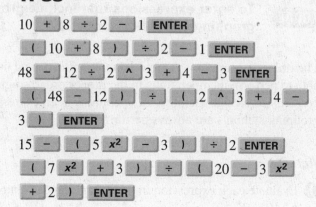

10 `+` 8 `÷` 2 `−` 1 `ENTER`

`(` 10 `+` 8 `)` `÷` 2 `−` 1 `ENTER`

48 `−` 12 `÷` 2 `^` 3 `+` 4 `−` 3 `ENTER`

`(` 48 `−` 12 `)` `÷` `(` 2 `^` 3 `+` 4 `−`

3 `)` `ENTER`

15 `−` `(` 5 `x²` `−` 3 `)` `÷` 2 `ENTER`

`(` 7 `x²` `+` 3 `)` `÷` `(` 20 `−` 3 `x²`

`+` 2 `)` `ENTER`

SHARP EL-9600c

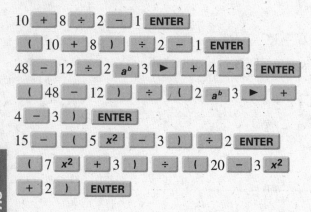

10 `+` 8 `÷` 2 `−` 1 `ENTER`

`(` 10 `+` 8 `)` `÷` 2 `−` 1 `ENTER`

48 `−` 12 `÷` 2 `aᵇ` 3 `▶` `+` 4 `−` 3 `ENTER`

`(` 48 `−` 12 `)` `÷` `(` 2 `aᵇ` 3 `▶` `+`

4 `−` 3 `)` `ENTER`

15 `−` `(` 5 `x²` `−` 3 `)` `÷` 2 `ENTER`

`(` 7 `x²` `+` 3 `)` `÷` `(` 20 `−` 3 `x²`

`+` 2 `)` `ENTER`

CASIO CFX-9850Gᴀ PLUS

From the main menu, choose RUN.

10 `+` 8 `÷` 2 `−` 1 `EXE`

`(` 10 `+` 8 `)` `÷` 2 `−` 1 `EXE`

48 `−` 12 `÷` 2 `^` 3 `+` 4 `−` 3 `EXE`

`(` 48 `−` 12 `)` `÷` `(` 2 `^` 3 `+` 4 `−`

3 `)` `EXE`

15 `−` `(` 5 `x²` `−` 3 `)` `÷` 2 `EXE`

`(` 7 `x²` `+` 3 `)` `÷` `(` 20 `−` 3 `x²`

`+` 2 `)` `EXE`

Algebra 1
Chapter 1 Resource Book

NAME _____ DATE _____

Practice A

For use with pages 15–21

Name the operation that would be performed first.

1. $16 - 8 + 4$ **2.** $2 \cdot 9 \div 3$ **3.** $10 \div (8 - 6)$

4. $16 \div 8 \cdot 2$ **5.** $16 + 8 \div 4 - 2$ **6.** $5 \cdot (4 + 2)^2$

Evaluate the expression for the given value of the variable.

7. $3 + 2x^2$ when $x = 2$ **8.** $30 - 3x^2$ when $x = 3$ **9.** $2y^2 + 5$ when $y = 3$

10. $4 \cdot 2a^3$ when $a = 10$ **11.** $2y^2 - 3y$ when $y = 4$ **12.** $6 + x^3$ when $x = 2$

13. $a^2 \div 5 + 3$ when $a = 5$ **14.** $32 - \dfrac{24}{n}$ when $n = 6$ **15.** $\dfrac{x}{3} + x^2 - 2$ when $x = 9$

Evaluate the expression.

16. $5 + 2 - 3$ **17.** $12 - 6 + 1$ **18.** $10 \cdot 2 \div 4$

19. $4 + 3 \cdot 2$ **20.** $8 \cdot 3 - 10$ **21.** $5 - 14 \div 7$

22. $2 + 36 \div 4$ **23.** $10 \div 5 + 3 \cdot 2$ **24.** $4 - 20 \div 10 + 7$

25. $3 \cdot 2^2 + 1$ **26.** $2 \cdot 3^2 \div 3$ **27.** $4(2 + 3) - 18$

Two calculators were used to evaluate the expression. They gave different results. Which calculator used the correct order of operations?

28. $12 \boxed{-} 4 \boxed{\times} 2 \boxed{+} 1 \boxed{\text{ENTER}}$
Calculator 1: 5   Calculator 2: 17

29. $5 \boxed{\times} 3 \boxed{-} 4 \boxed{\times} 2 \boxed{\text{ENTER}}$
Calculator 1: 7   Calculator 2: 22

30. $2 \boxed{\times} 6 \boxed{+} 3 \boxed{\div} 3 \boxed{\text{ENTER}}$
Calculator 1: 5   Calculator 2: 13

31. $10 \boxed{-} 5 \boxed{\times} 4 \boxed{\div} 10 \boxed{\text{ENTER}}$
Calculator 1: 8   Calculator 2: 2

32. *Shotput*   During a track meet, Kelly throws the shotput 36 feet, 35 feet, and 37 feet. Write an expression that represents the length of his average throw in feet. Evaluate the expression.

33. *Sales Tax*   You want to buy a newly released CD. The CD costs $17 plus 6% tax. (6% = 0.06) Write an expression that represents how much money in dollars you need to buy the CD. Evaluate the expression.

34. *Fish Tank*   You have two fish tanks. The first has a length of 4 feet, a width of 2 feet, and a height of 3 feet. The second has length 3 feet, width 2 feet, and height 1 foot. Write an expression that represents the total amount of water in cubic feet both tanks will hold. Evaluate the expression.

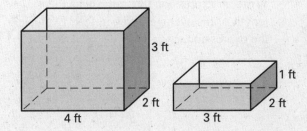

Practice B

For use with pages 15–21

Evaluate the expression. Then simplify the answer.

1. $\dfrac{6 + 4}{2^4 + 4 \div 2}$

2. $\dfrac{2^3 \cdot 5}{3^2 - 3 \cdot 2}$

3. $\dfrac{3^4 - 4^3}{2 \cdot 8 + 1}$

Evaluate the expression for the given value of the variable.

4. $x^5 + 7$ when $x = 3$

5. $3y^3 \div 4$ when $y = 2$

6. $40 - 2a^2$ when $a = 4$

7. $16 + 9b$ when $b = 5$

8. $x^2 - 3x$ when $x = 7$

9. $2 \cdot 8t^2$ when $t = 4$

10. $\dfrac{36}{a} + a$ when $a = 9$

11. $\dfrac{3}{5} \cdot y \div \dfrac{1}{10}$ when $y = \dfrac{3}{4}$

12. $\dfrac{1}{2} \cdot \dfrac{48}{b} + 7$ when $b = 6$

13. $\dfrac{x}{12} - \dfrac{2}{3} + 2$ when $x = 11$

Evaluate the expression.

14. $6 \div 3 \cdot 8$

15. $\dfrac{3}{5} - 2 \div 10$

16. $5^2 - 14 \div 7$

17. $2 + 3.6 \div 0.4$

18. $10 \div 5 + 3 \cdot 2$

19. $4 - 20 \div 10 + 7$

20. $\dfrac{3}{4} \cdot 2^2 + 1$

21. $\dfrac{2}{3} \cdot 3^2 \div 3$

22. $12(2 + 0.5) - 18$

23. $\dfrac{3}{5}(4 \cdot 10) - 6$

24. $[(7 - 5)^5 \div 8] - 4$

25. $2[(9 - 8)^2 + (12 - 5)^2]$

Two calculators were used to evaluate the expression. They gave different results. Which calculator used the correct order of operations?

26. $7\ \boxed{+}\ 3\ \boxed{\times}\ 6\ \boxed{-}\ 12\ \boxed{\text{ENTER}}$

Calculator 1: 13   Calculator 2: 48

27. $15\ \boxed{\div}\ 3\ \boxed{+}\ 4\ \boxed{\times}\ 2\ \boxed{\text{ENTER}}$

Calculator 1: 18   Calculator 2: 13

28. $7\ \boxed{+}\ 7\ \boxed{\div}\ 7\ \boxed{\times}\ 7\ \boxed{\div}\ 7\ \boxed{\text{ENTER}}$

Calculator 1: 2   Calculator 2: 8

29. $3\ \boxed{+}\ 2\ \boxed{\wedge}\ 3\ \boxed{-}\ 5\ \boxed{\text{ENTER}}$

Calculator 1: 6   Calculator 2: 120

30. *Shotput*   During a track meet, Kelly throws the shotput 51 feet, 50 feet, and 58 feet. Write an expression that represents the length of his average throw in feet. Evaluate the expression.

31. *Sales Tax*   You want to buy a newly released CD. The CD costs $14.95 plus 6% tax. (6% = 0.06) Write an expression that represents how much money in dollars you need to buy the CD. Evaluate the expression. Round to the nearest cent.

NAME _____ DATE _____

Reteaching with Practice

For use with pages 15–21

GOAL **Use the established order of operations.**

VOCABULARY

An established **order of operations** is used to evaluate an expression involving more than one operation.

EXAMPLE 1 *Evaluate Expressions Without Grouping Symbols*

a. Evaluate $5x^2 - 6$ when $x = 3$. Use the order of operations.

b. Evaluate $7 + 15 \div 3 - 4$. Use the order of operatons.

SOLUTION

a. $5x^2 - 6 = 5 \cdot 3^2 - 6$	Substitute 3 for x.
$= 5 \cdot 9 - 6$	Evaluate power.
$= 45 - 6$	Evaluate product.
$= 39$	Evaluate difference.
b. $7 + 15 \div 3 - 4 = 7 + (15 \div 3) - 4$	Divide first.
$= 7 + 5 - 4$	Evaluate quotient.
$= 12 - 4$	Work from left to right.
$= 8$	Evaluate difference.

Exercises for Example 1

Evaluate the expression.

1. $4 \cdot 3 + 8 \div 2$ **2.** $24 \div 6 \cdot 2$ **3.** $21 - 5 \cdot 2$

EXAMPLE 2 *Evaluate Expressions With Grouping Symbols*

Evaluate $24 \div (6 \cdot 2)$. Use the order of operations.

SOLUTION

$24 \div (6 \cdot 2) = 24 \div 12$	Simplify $6 \cdot 2$.
$= 2$	Evaluate the quotient.

Exercises for Example 2

Evaluate the expression.

4. $(6 - 2)^2 - 1$ **5.** $30 \div (1 + 4) + 2$ **6.** $(8 + 4) \div (1 + 2) + 1$

7. $6 - (2^2 - 1)$ **8.** $(30 \div 1) + (4 + 2)$ **9.** $8 + 4 \div (1 + 2 + 1)$

NAME _____ DATE _____

Reteaching with Practice

For use with pages 15–21

EXAMPLE 3 *Calculate Family Admission Prices*

Use the table below which shows admission prices for a theme park. Suppose a family of 2 adults and 3 children go to the park. The children's ages are 6 years, 8 years, and 13 years.

a. Write an expression that represents the admission price for the family.

b. Use a calculator to evaluate the expression.

Theme Park Admission Prices	
Age	*Admission Price*
Adults	$34.00
Children (3–9 years)	$21.00
Children (2 years and under)	free

SOLUTION

a. The admission price for the child who is 13 years old is $34, the adult price. The family must buy 3 adult tickets and 2 children's tickets. An expression that represents the admission price for the family is $3(34) + 2(21)$.

b. If your calculator uses the established order of operations, the following keystroke sequence gives the result 144.

3 ⨉ 34 + 2 ⨉ 21 **ENTER**

The admission price for the family is $144.

Exercise for Example 3

10. Rework Example 3 for a family of 2 adults and 4 children. The children's ages are 2 years, 4 years, 10 years, and 12 years.

NAME _____ DATE _____

Quick Catch-Up for Absent Students

For use with pages 15–21

The items checked below were covered in class on (date missed) _____

Lesson 1.3: Order of Operations

_____ **Goal:** Use the established order of operations.

Material Covered:

_____ Example 1: Evaluate Without Grouping Symbols

_____ Student Help: Study Tip

_____ Example 2: Use the Left-to-Right Rule

_____ Student Help: Skills Review

_____ Example 3: Expressions with Fraction Bars

_____ Example 4: Using a Calculator

_____ Example 5: Evaluate a Real-Life Expression

Vocabulary:

order of operations, p. 15   left-to-right rule p.16

_____ Other (specify) _____

_____

Homework and Additional Learning Support

_____ Textbook (specify) _pp. 18–21_____

_____

_____ *Reteaching with Practice* worksheet (specify exercises)_____

_____ *Personal Student Tutor* for Lesson 1.3

NAME _____ DATE _____

Real-Life Application: When Will I Ever Use This?

For use with pages 15–21

Back to School Shopping

You and a friend have just finished shopping for "back to school" clothes. Each of you was allotted $200.00 to spend. The table below shows the items that you have selected together and their cost. Assume that there is no tax on the selected items. To make things fair, you have agreed to split the total cost of the items equally.

Item	Quantity	Cost (per item)	Discount (per item) (where applicable)
Backpack	2	$18.00	$4.50 off
Pants	2	$29.95	$5.99 off
T-shirt	4	$9.95	$1.99 off
Jeans	2	$32.00	$6.40 off
Shoes	2	$65.95	$26.38 off
Shorts	2	$12.95	$2.59 off
3-Pack of Socks	4	$9.00	no discount
Sweatshirt	2	$25.00	no discount

1. Write an expression that represents the total amount that each of you spent after all discounts have been applied. Use grouping symbols to assure that the expression is evaluated correctly.

2. Evaluate the expression you wrote in Exercise 1.

3. Write an expression that shows how much money you and your friend saved after the discounts have been applied.

4. Evaluate the expression you wrote in Exercise 3.

5. How much money do you have left? Explain in writing what additional items you could purchase with your extra money.

NAME _____ DATE _____

Challenge: Skills and Applications

For use with pages 15–21

For Exercises 1–2, evaluate each expression.

1. $\dfrac{7^2 + (12 - 3)^2}{[4^2 - (5 + 1)]^2}$

2. $\dfrac{3.75 - 0.6(2.4)}{(5 - 2) \div 10}$

For Exercises 3–4, use the following information.

A clothing store has a sale with all items marked 25% off. The sales tax where the store is located is 5.5%.

3. Find the actual cost of each item in the table.

Item	*Original Price*	*Actual Cost*
Shirt	$35.00	
Blouse	$42.00	
Shoes	$48.00	
Belt	$22.00	

4. Write an expression for the actual cost of an item from Exercise 3 if the original price is p.

To find the volume of a right prism, multiply the area of the base times the height. For Exercises 5–7, write an expression with grouping symbols for the volume of a right prism with the height h and the base shown.

5.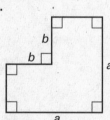

6.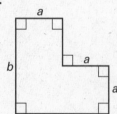

7.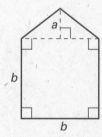

NAME _____ DATE _____

Quiz 1

For use after Lessons 1.1–1.3

1. Evaluate $5y$ when $y = 4$. *(Lesson 1.1)*

2. Evaluate $\dfrac{n}{24}$ when $n = 96$. *(Lesson 1.1)*

3. Find the distance traveled by a car moving at an average speed of 45 miles per hour for 2 hours. *(Lesson 1.1)*

4. Evaluate x^4 when $x = 5$. *(Lesson 1.2)*

5. Evaluate $(x - y)^2$ when $x = 10$ and $y = 3$. *(Lesson 1.2)*

6. Find the volume of a cube measuring 6 feet on each side. *(Lesson 1.2)*

7. Evaluate the expression $\dfrac{4 \cdot 3 + 6}{(3 + 2) - 4}$. *(Lesson 1.3)*

8. Evaluate $(5 + 2)^3 - 10$. *(Lesson 1.3)*

9. Evaluate $z^2 - 7 \cdot 8$ when $z = 10$. *(Lesson 1.3)*

Answers

1. _____
2. _____
3. _____
4. _____
5. _____
6. _____
7. _____
8. _____
9. _____

TEACHER'S NAME _____ CLASS _____ ROOM _____ DATE _____

Lesson Plan

1-day lesson (See *Pacing the Chapter,* TE page 1A)   **For use with pages 22–29**

GOAL **Check solutions of equations and inequalities.**

State/Local Objectives _____

_____

✓ **Check the items you wish to use for this lesson.**

STARTING OPTIONS

____ Homework Check (1.3): TE page 18; Answer Transparencies

____ Homework Quiz (1.3): TE page 21, CRB page 49, or Transparencies

____ Warm-Up: CRB page 49 or Transparencies

TEACHING OPTIONS

____ Developing Concepts: SE pages 22–23; CRB page 50 (Activity Support Master)

____ Lesson Opener: CRB page 51 or Transparencies

____ Examples 1–5: SE pages 24–26

____ Extra Examples: TE pages 25–26 or Transparencies; Internet Help at www.mcdougallittell.com

____ Checkpoint Exercises: SE pages 25–26

____ Concept Check: TE page 26

____ Guided Practice Exercises: SE page 27

APPLY/HOMEWORK

Homework Assignment

____ Transitional: SRH p. 771 Exs. 1–12; pp. 27–29, Exs. 26–29, 34–36, 49–52, 57, 61–63, 68–70, 83–88

____ Average: pp. 27–29, Exs. 30, 31, 37–42, 53, 54, 57, 58, 61–63, 71–76, 80–85

____ Advanced: pp. 27–29, Exs. 32, 33, 43–48, 55, 56, 59, 63, 66, 67, 71–73, 80–82, 89–91; EC: CRB p. 58

Reteaching the Lesson

____ Practice Masters: CRB pages 52–53 (Level A, Level B)

____ Reteaching with Practice: CRB pages 54–55 or Practice Workbook with Examples; Resources in Spanish

____ Personal Student Tutor: CD-ROM

Extending the Lesson

____ Interdisciplinary/Real-Life Applications: CRB page 57

____ Challenge: CRB page 58

ASSESSMENT OPTIONS

____ Daily Quiz (1.4): TE page 29, CRB page 61, or Transparencies

____ Standardized Test Practice: SE page 29; STP Workbook; Transparencies

Notes _____

_____

_____

LESSON 1.4

Lesson Plan for Block Scheduling

Half-block lesson (See *Pacing the Chapter,* TE page 1A)   **For use with pages 22–29**

GOAL   **Check solutions of equations and inequalities.**

State/Local Objectives _____

_____

_____

✓ Check the items you wish to use for this lesson.

CHAPTER PACING GUIDE	
Day	Lesson
1	1.1 (all); 1.2 (all)
2	1.3 (all); **1.4 (all)**
3	1.5 (all); 1.6 (all)
4	1.7 (all); 1.8 (all)
5	Ch. 1 Review and Assess

STARTING OPTIONS

____ Homework Check (1.3): TE page 18; Answer Transparencies
____ Homework Quiz (1.3): TE page 21,
       CRB page 49, or Transparencies
____ Warm-Up: CRB page 49 or Transparencies

TEACHING OPTIONS

____ Developing Concepts: SE pages 22–23; CRB page 50 (Activity Support Master)
____ Lesson Opener: CRB page 51 or Transparencies
____ Examples 1–5: SE pages 24–26
____ Extra Examples: TE pages 25–26 or Transparencies; Internet Help
       at www.mcdougallittell.com
____ Checkpoint Exercises: SE pages 25–26
____ Concept Check: TE page 26
____ Guided Practice Exercises: SE page 27

APPLY/HOMEWORK

Homework Assignment (See also the assignment for Lesson 1.3.)

____ Block Schedule: pp. 27–29, Exs. 30, 31, 37–42, 53, 54, 57, 58, 61–63, 71–76, 80–85

Reteaching the Lesson

____ Practice Masters: CRB pages 52–53 (Level A, Level B)
____ Reteaching with Practice: CRB pages 54–55 or Practice Workbook with Examples;
       Resources in Spanish
____ Personal Student Tutor: CD-ROM

Extending the Lesson

____ Interdisciplinary/Real-Life Applications: CRB page 57
____ Challenge: CRB page 58

ASSESSMENT OPTIONS

____ Daily Quiz (1.4): TE page 29, CRB page 61, or Transparencies
____ Standardized Test Practice: SE page 29; STP Workbook; Transparencies

Notes _____

_____

_____

Lesson 1.4

WARM-UP EXERCISES

For use before Lesson 1.4, pages 22–29

Evaluate the expression.

1. $\dfrac{x}{3} - 6$ when $x = 24$

2. $3x^3$ when $x = 4$

3. $3k - k$ when $k = 12$

4. Tell if the statement is true or false.

    a. $7x$ equals 12 when $x = 2$.

    b. $6 + b$ equals 8 when $b = 2$.

DAILY HOMEWORK QUIZ

For use after Lesson 1.3, pages 15–21

Evaluate the expression. Simplify the answer when necessary.

1. $12 \div 3 + 2 \cdot 8$

2. $\dfrac{8 \cdot 2 + 5}{12 + 2^2 - 9}$

Evaluate the expression when $x = 5$.

3. $4x^2 - 16$

4. $3(x + 6) - 8$

5. A T-shirt company charges \$15 per shirt and offers a \$200 rebate for orders greater than 50 shirts. Write an expression that represents the cost for an order of 75 shirts. Evaluate the expression.

NAME _____ DATE _____

Activity Support Master

For use with pages 22 and 23

Step 1

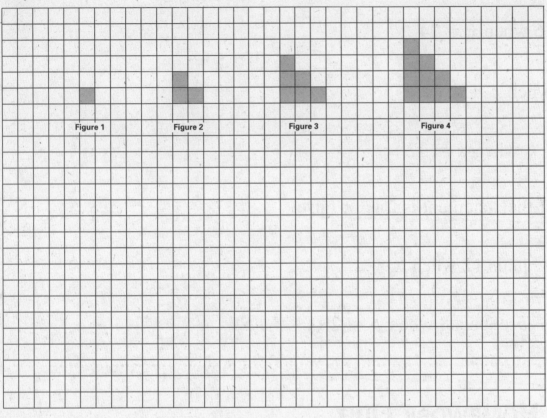

Figure 1   Figure 2   Figure 3   Figure 4

Step 2

Figure	1	2	3	4	5	6
Perimeter	4	8	12	16		
Pattern	4 · 1	4 · 2	4 · 3	4 · 4		

Step 3

Perimeter of 10th figure _____

Application Lesson Opener

For use with pages 22–29

1. You earned $2 more this week than you earned the week before. You earned $36 this week. How much did you earn last week? You can use an equation to solve this problem. Let x represent the amount you earned last week. The equation $x + 2 = 36$ models this situation.

   a. How can you find a value for the variable that makes this equation true?

   b. How do you know that your value makes the equation true?

2. You and two friends are sharing a pizza. You each get 4 slices of pizza. How many slices of pizza are there in all? You can use an equation to solve this problem. Let x represent the number of slices of pizza in all. The equation $\dfrac{x}{3} = 4$ models this situation.

   a. How can you find a value for the variable that makes this equation true?

   b. How do you know that your value makes the equation true?

3. You bought some CDs and a cassette tape at a clearance sale. You paid $8 for each CD and $5 for the cassette tape. Your total purchase before tax came to $29. How many CDs did you buy? You can use an equation to solve this problem. Let x represent the number of CDs you bought. The equation $8x + 5 = 29$ models this situation.

   a. How can you find a value for the variable that makes this equation true?

   b. How do you know that your value makes the equation true?

Lesson 1.4

NAME _____ DATE _____

Practice A
For use with pages 22–29

Decide whether the following is an *expression,* an *equation,* or an *inequality.*

1. $5x + 9 = 12$ **2.** $7x - 2$ **3.** $3x - 2 \geq 12$

4. $5.5 = 3x - 9$ **5.** $19 < 4a$ **6.** $89 - 2a - 39$

Check whether the given number is a solution of the equation.

7. $2x + 3 = 7; 4$ **8.** $4x + 2 = 10; 1$ **9.** $3x - 5 = 1; 2$

10. $6 = 2x - 8; 6$ **11.** $17 - 4a = 13; 1$ **12.** $5a - 3 = 2a; 3$

13. $4y - 6 = 2y; 3$ **14.** $y + 3y = 2y + 4; 3$ **15.** $5x + 3 = x + 7; 1$

16. $4 + \dfrac{m}{3} = 12; 9$ **17.** $x^2 - 6 = 30; 6$ **18.** $3n + \dfrac{10}{n} = 17; 5$

Write a question that could be used to solve the equation. Then use mental math to solve the equation.

19. $x - 5 = 3$ **20.** $x + 2 = 6$ **21.** $4 + x = 6$

22. $12 - x = 5$ **23.** $4t = 20$ **24.** $30 \div x = 6$

25. $\dfrac{x}{2} = 3$ **26.** $\dfrac{18}{m} = 3$ **27.** $x^3 = 8$

Check whether the given number is a solution of the inequality.

28. $x - 5 \leq 7; 9$ **29.** $x + 3 > 7; 4$ **30.** $3 + x < 8; 5$

31. $10 - x > 2; 9$ **32.** $2x + 1 \geq 10; 6$ **33.** $4x - 3 \leq 5; 2$

34. $3 - \dfrac{x}{2} > 0; 2$ **35.** $6x + 1 \geq 8x - 7; 2$ **36.** $5x + 1 \geq x - 3; 4$

37. *Locker Installation* Suppose your school is replacing some of its lockers. When the old lockers are removed there is a space 144 inches long. Each new locker has a width of 8 inches. You want to know how many new lockers can be installed. You write the inequality $8x \leq 144$ to model the situation. What do the 8, x, and 144 represent? Use mental math to solve the inequality.

38. *Statue of Liberty* The Statue of Liberty's torch has 14 lamps that give off 14,000 watts of light. You want to know how many watts are given off in one lamp if all the lamps are identical. You write the equation $14x = 14,000$ to model the situation. What do the 14, x, and 14,000 represent? Use mental math to solve the equation.

NAME _____ DATE _____

Practice B

For use with pages 22–29

Check whether the given number is a solution of the equation.

1. $2x + 3 = 17; 7$

2. $10 = 2x - 8; 6$

3. $4 + x^2 = 10; 3$

4. $x^2 - 19 = 30; 7$

5. $3x - 5 = x - 1; 2$

6. $17 - 2a^3 = 11; 1$

7. $5a - 3 = 3 + 2a; 2$

8. $y + 3y = 2y + 6; 3$

9. $5x + 3x = 30 - 2x; 5$

10. $\frac{m}{3} - 4 = 12; 24$

11. $9 + \frac{y}{2} = 2y; 6$

12. $3n - n = \frac{10}{n}; 5$

Write a question that could be used to solve the equation. Then use mental math to solve the equation.

13. $x - 5 = 8$

14. $x + 2 = 12$

15. $6x - 9 = 9$

16. $24 - x = 15$

17. $4x = 32$

18. $3x + 7 = 22$

19. $\frac{x}{7} = 3$

20. $\frac{36}{m} = 3$

21. $x^3 = 64$

Check whether the given number is a solution of the inequality.

22. $x - 5 \le 7; 12$

23. $9 + x > 13; 4$

24. $4x + 1 < 8; 2$

25. $10 - x \ge x; 4$

26. $x^2 + 7 \ge 10; 2$

27. $\frac{4}{x} + 3 \le 5; 2$

28. $0 > \frac{x - 3}{6}; 9$

29. $6(x + 1) \ge 8x - 7; 2$

30. $\frac{10 + c}{c} < c - 3; 5$

Match the verbal sentence with its mathematical representation.

31. The sum of 20 and x is greater than 30.

32. The quotient of x and 20 is greater than or equal to 30.

33. The product of 30 and x is less than 10.

34. The square of x is equal to 36.

35. The difference of 4 and x is less than or equal to 30.

A. $x^2 = 36$

B. $20 + x > 30$

C. $4 - x \le 30$

D. $\frac{x}{20} \ge 30$

E. $30(x) < 10$

36. *Locker Installation* Suppose your school is replacing some of its lockers. When the old lockers are removed there is a space 200 inches long. Each new locker has a width of 8 inches. You want to know how many new lockers can be installed. You write the inequality $8x \le 200$ to model the situation. What do the 8, x, and 200 represent? Use mental math to solve the inequality.

37. *Budget* You are budgeting money to buy a video game system that costs $145 including tax. If you save $5 per week, when will you have enough money? You write the inequality $5n \ge 145$ to model the situation. What do the 5, n, and 145 represent? Use mental math to solve the inequality.

NAME _____ DATE _____

Reteaching with Practice

For use with pages 22–29

GOAL **Check solutions of equations and inequalities.**

VOCABULARY

An **equation** is a statement formed when an equal sign is placed between two expressions.

When the variable in an equation is replaced by a number and the resulting statement is true, the number is a **solution of the equation.**

Finding all the solutions of an equation is called **solving the equation.**

An **inequality** is a statement formed when an inequality symbol is placed between two expressions.

A **solution of an inequality** is a number that produces a true statement when it is substituted for the variable in the inequality.

EXAMPLE 1 *Check Possible Solutions of an Equation*

Check whether the numbers 2 and 4 are solutions of the equation
$2x + 3 = 11$

SOLUTION

To check the possible solutions, substitute them into the equation. If both sides of the equation have the same value, then the number is a solution.

x	$2x + 3 = 11$	Result	Conclusion
2	$2(2) + 3 \stackrel{?}{=} 11$	$7 \neq 11$	2 is not a solution
4	$2(4) + 3 \stackrel{?}{=} 11$	$11 = 11$	4 is a solution

The number 4 is a solution of $2x + 3 = 11$. The number 2 is not a solution.

Exercises for Example 1

Check whether the given number is a solution of the equation.

1. $5p - 2 = 12; 3$ **2.** $8 + 2y = 10; 3$ **3.** $3a + 2 = 14; 2$

4. $\dfrac{t}{4} - 3 = 0; 12$ **5.** $n + 4n = 20; 5$ **6.** $k + 7 = 3k + 1; 3$

Reteaching with Practice

For use with pages 22–29

EXAMPLE 2 *Use Mental Math to Solve an Equation*

Which question could be used to find the solution of the equation $x - 7 = 15$?

A. What number can be subtracted from 7 to get 15?

B. What number can 7 be subtracted from to get 15?

C. What number can 15 be subtracted from to get 7?

SOLUTION

Because 7 can be subtracted from 22 to get 15, question B could be used to solve the equation $x - 7 = 15$.

Exercises for Example 2

Write a question that could be used to solve the equation. Then use mental math to solve the equation.

7. $x + 7 = 21$ 8. $3f + 1 = 19$ 9. $a - 12 = 10$

10. $\dfrac{y}{3} = 11$ 11. $4j - 7 = 9$ 12. $\dfrac{b}{2} = 4$

EXAMPLE 3 *Check Solutions of Inequalities*

Decide whether 6 is a solution of the inequality.

a. $3 + w \geq 9$ **b.** $r + 4 > 11$

SOLUTION

	Inequality	Substitution	Result	Conclusion
a.	$3 + w \geq 9$	$3 + 6 \overset{?}{\geq} 9$	$9 \geq 9$	6 is a solution.
b.	$r + 4 > 11$	$6 + 4 \overset{?}{>} 11$	$10 \not> 11$	6 is not a solution.

Exercises for Example 3

Check whether the given number is a solution of the inequality.

13. $2f - 3 \geq 8; 5$ 14. $2h - 4 > 10; 3$ 15. $13x \leq 6x + 15; 2$

NAME _____ DATE _____

Quick Catch-Up for Absent Students

For use with pages 22–29

The items checked below were covered in class on (date missed) _____

Activity 1.4: Finding Patterns (p. 23)

____ **Goal:** Use algebraic expressions to describe a pattern.

Lesson 1.4: Equations and Inequalities

____ **Goal:** Check solutions of equations and inequalities.

Material Covered:

____ Student Help: Vocabulary Tip

____ Example 1: Check Possible Solutions

____ Example 2: Solve Equations with Mental Math

____ Example 3: Use Mental Math to Solve a Real-Life Equation

____ Student Help: Study Tip

____ Example 4: Check Solutions of Inequalities

____ Example 5: Check Solutions in Real-Life

Vocabulary:

equation, p. 24   solving the equation, p. 25

solution of an equation, p. 24   inequality, p. 26

solution of an inequality, p. 26

____ Other (specify) _____

_____

Homework and Additional Learning Support

____ Textbook (specify) pp. 27–29 _____

_____

____ Internet: Extra Examples at www.mcdougalllittell.com

____ *Reteaching with Practice* worksheet (specify exercises) _____

____ *Personal Student Tutor* for Lesson 1.4

NAME _____ DATE _____

Interdisciplinary Application

For use with pages 22–29

Drums

MUSIC A drum is a type of membranophone, an instrument in which sound is made by vibration of a stretched membrane or skin. The earliest evidence of drums dates back about 8000 years ago. Drums were depicted in Egyptian art 6000 years ago and in Mesopotamian art 5000 years ago. Though many drum frames have been discovered, few complete drums have survived due to the perishable nature of the animal membranes that were used. They have been included in Western orchestras since the 1600s.

Drums are generally categorized by their shape (tubular, vessel, frame) or by their method of sounding (friction). Tubular drums are shaped like their name implies and include the conical, barrel, cylindrical, waisted, goblet, footed, and long drums. Vessel drums have a pot-shaped body and single head, generally referred to as kettledrums. Frame drums have their skin stretched over the frame and include the tambourine.

The skins of the drums can be attached using glue, nails, pegs, or laces in a variety of patterns, or a hoop. Each method produces a slightly different tone when sounded. Larger drums generally have a lower pitch, and those with tighter membranes have a higher pitch. Drums are sounded by creating vibrations on the skins using hands, sticks, beaters, and wire brushes. Clapper drums are struck with heads when shaken, and friction drums are sounded by rubbing. Present-day drums can be tuned to have specific pitches, and most use synthetic materials to replace the animal membranes. Most instrumental groups will include a bass drum, a gong, a timpani or kettledrum, a snare or side drum, a tenor, and a tambourine.

In Exercises 1–4, use the following information.

You have just recently started playing the drums, with the hope that you can be in the school's jazz band next year. The band director says that you need at least 200 hours of practice time before you audition in the fall.

1. You have twelve weeks during the summer to start playing the drums. What is the least number of hours that you will have to practice each week and still be able to audition in the fall? Write an inequality to model this situation.

2. What do each of the numbers and the variable represent in your inequality from Exercise 1?

3. Solve the inequality from Exercise 1. Round your answer to the nearest hour.

4. Does your answer seem to be reasonable? Why or why not?

Algebra 1
Chapter 1 Resource Book

57

NAME _____ DATE _____

Challenge: Skills and Applications

For use with pages 22–29

Check whether the given number is a solution of the equation or inequality.

1. $\dfrac{28 - x^2}{3x} = 2 + \dfrac{x + 2}{2x}$; 4

2. $\dfrac{3}{1 - x} - \dfrac{9}{2x} = 6x$; 0.75

3. $\dfrac{8}{2x + 2} = \dfrac{1}{x - \frac{1}{8}}$; $\dfrac{1}{2}$

4. $3 - x^2 = \dfrac{x}{2}$; $\dfrac{3}{2}$

5. $(2x - 7)4 + 3 \geq \dfrac{5x}{3}$; 4

6. $5 + 2x^2 + 6 < 2^x$; 3

7. $\dfrac{7 - 2x}{2x + 1} > \dfrac{x^2 + 1}{2^{x+1}}$; 2

8. $6 - 4x^2 \leq 6 - 4x \div 2 \cdot 5 + 3$; $\dfrac{1}{3}$

For Exercises 9–12, use the following information.

The total mass M (in kilograms) of a spacecraft that can be propelled by a magnetic sail around Earth is, in theory, given by

$$M = \dfrac{0.015m^2}{f}$$

where m is the mass (in kilometers) of the magnetic sail and f is the drag force (in Newtons) of the spacecraft.

A 4000-kilogram magnetic sail theoretically propels a 50,000-kilogram spacecraft in orbit around Earth.

9. Write an equation that models the situation.

10. Is 5 Newtons a solution to the equation from Exercise 9?

11. Is 4 Newtons a solution to the equation from Exercise 9?

12. Use mental math to estimate the solution. Justify your answer.

TEACHER'S NAME _____ CLASS _____ ROOM _____ DATE _____

Lesson Plan

1-day lesson (See *Pacing the Chapter,* TE page 1A)   **For use with pages 30–35**

GOAL **Translate words into mathematical symbols.**

State/Local Objectives _____

_____

✓ Check the items you wish to use for this lesson.

STARTING OPTIONS
____ Homework Check (1.4): TE page 27; Answer Transparencies
____ Homework Quiz (1.4): TE page 29, CRB page 61, or Transparencies
____ Warm-Up: CRB page 61 or Transparencies

TEACHING OPTIONS
____ Lesson Opener: CRB page 62 or Transparencies
____ Examples 1–6: SE pages 30–32
____ Extra Examples: TE pages 31–32 or Transparencies
____ Checkpoint Exercises: SE pages 30–32
____ Concept Check: TE page 32
____ Guided Practice Exercises: SE page 33

APPLY/HOMEWORK
Homework Assignment
____ Transitional: SRH p. 760 Exs. 16–21; pp. 33–35 Exs. 10–13, 20–27, 32, 33, 40–50, Quiz 2
____ Average: pp. 33–35 Exs. 14–17, 28–31, 34, 35, 37, 38, 40–46, 48–54 even, Quiz 2
____ Advanced: pp. 33–35 Exs. 16–19, 24–38 even, 39–42*, 51–54, Quiz 2; EC: CRB p. 69

Reteaching the Lesson
____ Practice Masters: CRB pages 63–64 (Level A, Level B)
____ Reteaching with Practice: CRB pages 65–66 or Practice Workbook with Examples; Resources in Spanish
____ Personal Student Tutor: CD-ROM

Extending the Lesson
____ Interdisciplinary/Real-Life Applications: CRB page 68
____ Challenge: CRB page 69

ASSESSMENT OPTIONS
____ Daily Quiz (1.5): TE page 35, CRB page 73, or Transparencies
____ Standardized Test Practice: SE page 35; STP Workbook; Transparencies
____ Quiz 1.4–1.5: SE page 35; CRB page 70; Resources in Spanish

Notes _____

_____

_____

TEACHER'S NAME _____ CLASS _____ ROOM _____ DATE _____

Lesson Plan for Block Scheduling

Half-block lesson (See *Pacing the Chapter,* TE page 1A)   For use with pages 30–35

GOAL   **Translate words into mathematical symbols.**

State/Local Objectives _____

_____

_____

✓ **Check the items you wish to use for this lesson.**

CHAPTER PACING GUIDE	
Day	Lesson
1	1.1 (all); 1.2 (all)
2	1.3 (all); 1.4 (all)
3	**1.5 (all)**; 1.6 (all)
4	1.7 (all); 1.8 (all)
5	Ch. 1 Review and Assess

STARTING OPTIONS
____ Homework Check (1.4): TE page 27; Answer Transparencies
____ Homework Quiz (1.4): TE page 29, CRB page 61, or Transparencies
____ Warm-Up: CRB page 61 or Transparencies

TEACHING OPTIONS
____ Lesson Opener: CRB page 62 or Transparencies
____ Examples 1–6: SE pages 30–32
____ Extra Examples: TE pages 31–32 or Transparencies
____ Checkpoint Exercises: SE pages 30–32
____ Concept Check: TE page 32
____ Guided Practice Exercises: SE page 33

APPLY/HOMEWORK
Homework Assignment (See also the assignment for Lesson 1.6.)
____ Block Schedule: pp. 33–35, Exs. 14–17, 28–31, 34, 35, 37, 38, 40–46, 48–54 even, Quiz 2

Reteaching the Lesson
____ Practice Masters: CRB pages 63–64 (Level A, Level B)
____ Reteaching with Practice: CRB pages 65–66 or Practice Workbook with Examples; Resources in Spanish
____ Personal Student Tutor: CD-ROM

Extending the Lesson
____ Interdisciplinary/Real-Life Applications: CRB page 68
____ Challenge: CRB page 69

ASSESSMENT OPTIONS
____ Daily Quiz (1.5): TE page 35, CRB page 73, or Transparencies
____ Standardized Test Practice: SE page 35; STP Workbook; Transparencies
____ Quiz 1.4–1.5: SE page 35; CRB page 70; Resources in Spanish

Notes _____

_____

_____

LESSON
1.5

NAME _____ DATE _____

Available as
a transparency

Lesson 1.5

WARM-UP EXERCISES

For use before Lesson 1.5, pages 30–35

Match the equation with the question that can be used to solve it.

1. $x - 3 = 9$

2. $3x = 9$

3. $9 - x = 3$

A. 3 times what number gives 9?

B. 9 minus what number gives 3?

C. What number minus 3 gives 9?

DAILY HOMEWORK QUIZ

For use after Lesson 1.4, pages 22–29

Check to see if the given value of the variable is or is not a solution of the equation or inequality.

1. $3x - 4 = 16; x = 7$

2. $y + 6 < 11; y = 4$

Use mental math to solve the equation.

3. $n - 5 = 17$

4. $3y = 36$

5. You want to jog at least 9 miles a week by jogging 2 miles per day. Use the inequality $2d > 9$ to find how many days you must jog each week to meet your goal.

Application Lesson Opener

For use with pages 30–35

Choose the equation that can be used to solve the problem. Explain why your choice is correct.

1. You are buying a computer system. The total system includes a CPU, a monitor, and a printer. The total cost of the system is $1300. If the CPU costs $895 and the monitor costs $225, what is the cost of the printer? Let x represent the cost of the printer.

   A. $1300 + x = 225 + 895$

   B. $x = 1300 + 225 + 895$

   C. $x + 225 + 895 = 1300$

2. You spent $75 on a three-day trip. You spent $21 the first day and half of the remaining amount on each of the last two days. How much did you spend on each of the last two days? Let x represent the amount you spent on each of the last two days.

   A. $2x + 21 = 75$

   B. $x + 2 + 21 = 75$

   C. $\dfrac{x}{2} + 21 = 75$

3. Your basketball team scored 56 points in your last game. Your team scored 12 points at the free-throw line and 9 two-point baskets. The rest of the points were scored on three-point baskets. How many three-point baskets did your team score? Let x represent the number of three-point baskets scored.

   A. $56 - x = 12 + 9 + 2$

   B. $12 + 2(9) + x = 56$

   C. $3x + 2(9) + 12 = 56$

NAME _____ DATE _____

Practice A

For use with pages 30–35

Write the verbal phrase as a variable expression. Use *x* for the variable in your expression.

1. Three more than a number

2. Four less than a number

3. Difference of eight and a number

4. The sum of a number and one

5. Six times a given number

6. One half of a given number

7. A number divided by five

8. Seven more than twice a given number

9. Two less than a number, divided by nine

10. Two more than the product of ten and a number

11. The sum of a number and one, times three

12. The sum of a number and six, divided by two

Write the verbal sentence as an equation or an inequality.

13. Two more than a number x is ten.

14. The sum of a number y and four is 13.

15. Eight more than a number y is greater than or equal to nine.

16. The difference of a number a and two is seven.

17. Six less than a number z is less than 15.

18. Eleven minus a number b is two.

19. The product of two and a number x is 22.

20. Twelve is less than six times a number x.

21. One more than four times a number b is five.

22. The quotient of a number t and three is eight.

23. A number a divided by two is greater than five.

24. Four less than the product of six and a number a is eight.

Algebra 1
Chapter 1 Resource Book

63

LESSON

1.5

NAME _____ DATE _____

Practice B

For use with pages 30–35

Write the verbal phrase as a variable expression. Use *x* for the variable in your expression.

1. Four more than a number

2. Six less than a number

3. Difference of 12 and a number

4. The sum of a number and two

5. Five times a given number

6. One third of a given number

7. A number divided by eight

8. Nine more than twice a given number

9. Two less than a number, divided by three

10. Three more than the product of ten and a number

11. Five times the sum of a number and one

12. The sum of a number and five, divided by two

Write the verbal sentence as an equation or an inequality.

13. Seven more than a number *x* is ten.

14. The sum of a number *y* and six is 13.

15. Eight more than a number *y* is greater than or equal to ten.

16. The difference of a number *a* and two is eight.

17. Six less than a number *z* is less than 21.

18. Thirteen minus a number *b* is two.

19. The product of 11 and a number *x* is 22.

20. Fourteen is less than seven times a number *x*.

21. One more than four times a number *b* is 17.

22. The quotient of a number *t* and three is nine.

23. A number *a* divided by two is greater than nine.

24. Three less than the product of six and a number *a* is nine.

Write the sentence as an equation. Let *x* represent the number Use mental math to solve the equation.

25. The sum of a number and 7 is 13.

26. The difference between 25 and a number is 15.

27. Six times a given number is 54.

28. A number divided by 8 is 9.

NAME _____ DATE _____

Reteaching with Practice

For use with pages 30–35

GOAL **Translate words into mathematical symbols.**

> **VOCABULARY**
>
> In order to solve real-life problems, you will **translate** words into mathematical symbols. In English, phrases are not complete sentences. In math, **phrases** are translated into variable expressions. Sentences are translated into equations or inequalities.

EXAMPLE 1 *Translate Addition and Subtraction Phrases*

Write the phrase as a variable expression. Let x represent the number.

Phrase	Translation
a. The sum of 3 and a number	$3 + x$
b. The difference between a number and 4	$x - 4$
c. 15 more than a number	$15 + x$
d. A number decreased by 7	$x - 7$

Exercises for Example 1

Write the phrase as a variable expression. Let x represent the number.

1. The sum of 1 and a number   **2.** 4 less than a number

3. 12 minus a number   **4.** A number plus 8

5. A number increased by 6   **6.** The difference between a number and 10

EXAMPLE 2 *Translate Multiplication and Division Phrases*

Write the phrase as a variable expression. Let n represent the number.

Phrase	Translation
a. 8 times a number	$8n$
b. The quotient of a number and 2	$\dfrac{n}{2}$
c. A number multiplied by 4	$4n$

Reteaching with Practice

For use with pages 30–35

Exercises for Example 2

Write the phrase as a variable expression. Let _n_ represent the number.

7. The product of 3 and a number

8. The quotient of a number and 4

9. 12 multiplied by a number

10. One half of a number

11. 9 divided by a number

EXAMPLE 3 ## Write and Solve an Equation

a. Translate into mathematical symbols: "The product of a number and 8 is 56." Let x represent the number.

b. Use mental math to solve your equation and check your solution.

SOLUTION

a. The equation is $8x = 56$.

b. Using mental math, the solution is $x = 7$.

$$8 \cdot x = 56$$
$$8 \cdot 7 = 56$$
$$56 = 56$$

Exercises for Example 3

Translate each sentence into mathematical symbols. Then use mental math to solve your equation. Let _x_ represent the number.

12. A number plus 6 is 15

13. The quotient of a number and 4 is 7.

14. A number increased by 8 is 15

15. A number multiplied by 3 is 33.

NAME _____ DATE _____

Quick Catch-Up for Absent Students

For use with pages 30–35

The items checked below were covered in class on (date missed) _____

Lesson 1.5: Translating Words into Mathematical Symbols

____ **Goal:** Translate words into mathematical symbols.

Material Covered:

    ____ Example 1: Translate Addition Phrases

    ____ Student Help: Reading Algebra

    ____ Example 2: Translate Subtraction Phrases

    ____ Student Help: Vocabulary Tip

    ____ Example 3: Translate Multiplication and Division Phrases

    ____ Student Help: Reading Algebra

    ____ Example 4: Translate Sentences

    ____ Student Help: Reading Algebra

    ____ Example 5: Write and Solve an Equation

    ____ Student Help: Skills Review

    ____ Example 6: Translate and Solve a Real-Life Problem

____ Other (specify) _____

_____

Homework and Additional Learning Support

    ____ Textbook (specify) _pp. 33–35_ _____

    _____

    ____ *Reteaching with Practice* worksheet (specify exercises) _____

    ____ *Personal Student Tutor* for Lesson 1.5

NAME _____ DATE _____

Real-Life Application: When Will I Ever Use This?

For use with pages 30–35

Taiwan Vacation

Window On China, or Xiao Ren Gwo (roughly translated as Little People World), is a special attraction located in Taiwan, Republic of China. This amusement park's actual location is 53 kilometers (or 33 miles) southwest of Taipei, the capital of Taiwan, in the city of Lungtan (near Taoyuan). Visitors who come to Window On China have an opportunity to view 100 miniature reproductions of many famous Chinese attractions from both mainland China and Taiwan, including the Great Wall of China and the Forbidden City. There are also additional miniature kingdoms containing many other famous structures from around the world. These reproductions are on a scale of 1:25, and are so detailed that it is difficult to distinguish between photos of the actual and miniature structures. Even the bushes and trees are created to scale. In addition to the miniature kingdoms, the park contains a classical Chinese garden, a restaurant, and an arcade area with games for kids.

The following prices for the park are in Taiwanese currency, New Taiwan dollars (NT$).

	Regular Prices	*Group Admissions*	*After 3:00 PM*
Adults	NT$590	NT$500	NT$420
Students	NT$500	NT$420	NT$420
Seniors/Children	NT$350	NT$300	NT$420

In Exercises 1–4, use the above pricing information as well as the following information.

A family decided to visit Window On China. Their total entrance fee was NT$8180. Be sure to use verbal models to write your algebraic equations.

1. If there were 7 adults and 3 children, how many students made the trip according to their total entrance fee? Assume the family arrived before 3:00 PM.

2. If they qualified for the group rate, how many students were present?

3. If the family consists of 7 adults and 7 students, and they do not qualify for the group rate, how much can they save by arriving after 3:00 PM?

4. If the exchange rate for New Taiwan dollars to U.S. dollars is 32:1, how much money in U.S. dollars did the family spend on entrance fees for Window On China? (Use the data from Exercise 1 and round your answer to the nearest cent.)

NAME _____ DATE _____

Challenge: Skills and Applications

For use with pages 30–35

For Exercises 1–10, use the following information.

Melika is putting up paper bordering on the walls of the three bedrooms of her house. She needs to measure each room to figure out how much she needs to buy for each room, and how much it will cost.

1. The first bedroom's length is 2 more feet than its width. If the width is *w,* what is the length?

2. The area of the first bedroom is 120 square feet. Write an equation, *but do not solve,* using the variable expressions in Exercise 1 to find the dimensions of the room.

3. The perimeter of the first bedroom is 44 feet. What are the dimensions?

4. The second bedroom's width is 3 feet less than its length. If the length is *l,* what is the width?

5. The area of the second bedroom is 180 square feet. Write an equation, but *do not solve,* using the variable expressions in Exercise 4 to find the dimensions of the room.

6. The perimeter of the second bedroom is 54 feet. What are the dimensions?

7. The third bedroom's length is $1\frac{1}{2}$ times the width. If the width is *d,* what is the length?

8. The area of the third bedroom is 216 square feet. Write an equation, but *do not solve,* using the variable expressions in Exercise 7 to find the dimensions of the room.

9. The perimeter of the third bedroom is 60 feet. What are the dimensions?

10. If the border for the first bedroom costs $1.95 per foot, the second bedroom is $1.45 per foot, the third bedroom is $1.75 per foot, and the sales tax rate is 5.5%, what will the total cost be?

For Exercises 11–13, use the following information.

When Scott looked at the temperature at 9:00 A.M., it had risen 7° from 7:30 A.M.

11. If the temperature was *t*° at 9:00 A.M., what was the temperature at 7:30 A.M.?

12. If it rises 12° (from the 9:00 temperature) by noon, what is the temperature then?

13. If, by 3:00 P.M., the temperature has risen 14° from the noon temperature to 86°, what was the temperature at 7:30 A.M.?

NAME _____ DATE _____

Quiz 2

For use after Lessons 1.4–1.5

1. Is $r = 4$ a solution of the equation $8 + r^2 = 16$? *(Lesson 1.4)*

2. Is $y = 8$ a solution of the inequality $2(5y - 4) \geq 65$?
   (Lesson 1.4)

3. Write a question that could be used to solve $3n + 2 = 14$.
   (Lesson 1.4)

4. Write a variable expression for "seven less than four times
   a number x." *(Lesson 1.5)*

**In Exercises 5 and 6, write an equation or inequality to model
the situation. (Lesson 1.5)**

5. The number h of hours worked this week plus the 20 hours you
   worked last week is less than 35.

6. The product of $20.00 and the number s of journal subscriptions is
   equal to $600.

Answers

1. _____

2. _____

3. _____

4. _____

5. _____

6. _____

Lesson Plan

1-day lesson (See *Pacing the Chapter*, TE page 1A)　　　　**For use with pages 36–41**

GOAL **Model and solve real-life problems.**

State/Local Objectives _____

_____

✓ Check the items you wish to use for this lesson.

STARTING OPTIONS
____ Homework Check (1.5): TE page 33; Answer Transparencies
____ Homework Quiz (1.5): TE page 35, CRB page 73, or Transparencies
____ Warm-Up: CRB page 73 or Transparencies

TEACHING OPTIONS
____ Lesson Opener: CRB page 74 or Transparencies
____ Examples 1–3: SE pages 36–38
____ Extra Examples: TE pages 37–38 or Transparencies
____ Checkpoint Exercises: SE pages 37–38
____ Concept Check: TE page 38
____ Guided Practice Exercises: SE page 39

APPLY/HOMEWORK
Homework Assignment
____ Transitional: pp. 39–41, Exs. 5–15, 22, 23–45 odd
____ Average: pp. 39–41, Exs. 5–10, 16–20, 22–32, 38–41
____ Advanced: pp. 39–41, Exs. 11–26*, 30–33, 42–44; EC: CRB p. 81

Reteaching the Lesson
____ Practice Masters: CRB pages 75–76 (Level A, Level B)
____ Reteaching with Practice: CRB pages 77–78 or Practice Workbook with Examples;
　　　Resources in Spanish
____ Personal Student Tutor: CD-ROM

Extending the Lesson
____ Interdisciplinary/Real-Life Applications: CRB page 80
____ Challenge: CRB page 81

ASSESSMENT OPTIONS
____ Daily Quiz (1.6): TE page 41, CRB page 84, or Transparencies
____ Standardized Test Practice: SE page 41; STP Workbook; Transparencies

Notes _____

_____

_____

TEACHER'S NAME _____ CLASS _____ ROOM _____ DATE _____

Lesson Plan for Block Scheduling

Half-block lesson (See *Pacing the Chapter*, TE page 1A.)　　　　For use with pages 36–41

GOAL　　Model and solve real-life problems.

State/Local Objectives _____

_____

_____

✓ **Check the items you wish to use for this lesson.**

STARTING OPTIONS

____ Homework Check (1.5): TE page 33; Answer Transparencies

____ Homework Quiz (1.5): TE page 35,
　　　CRB page 73, or Transparencies

____ Warm-Up: CRB page 73 or Transparencies

TEACHING OPTIONS

____ Lesson Opener: CRB page 74 or Transparencies

____ Examples 1–3: SE pages 36–38

____ Extra Examples: TE pages 37–38 or Transparencies

____ Checkpoint Exercises: SE pages 37–38

____ Concept Check: TE page 38

____ Guided Practice Exercises: SE page 39

APPLY/HOMEWORK

Homework Assignment (See also the assignment for Lesson 1.5.)

____ Block Schedule:　　pp. 39–41, Exs. 5–10, 16–20, 22–25, 28–31, 38–45

Reteaching the Lesson

____ Practice Masters: CRB pages 75–76　(Level A, Level B)

____ Reteaching with Practice: CRB pages 77–78 or Practice Workbook with
　　　Examples; Resources in Spanish

____ Personal Student Tutor: CD-ROM

Extending the Lesson

____ Interdisciplinary/Real-Life Applications: CRB page 80

____ Challenge: CRB page 81

ASSESSMENT OPTIONS

____ Daily Quiz (1.6): TE page 41, CRB page 84, or Transparencies

____ Standardized Test Practice: SE page 41; STP Workbook; Transparencies

Notes _____

_____

_____

CHAPTER PACING GUIDE	
Day	Lesson
1	1.1 (all); 1.2 (all)
2	1.3 (all); 1.4 (all)
3	1.5 (all); **1.6 (all)**
4	1.7 (all); 1.8 (all)
5	Ch. 1 Review and Assess

WARM-UP EXERCISES

For use before Lesson 1.6, pages 36–41

Solve the equation using mental math.

1. $3x = 6$

2. $x - 2 = 8$

3. $\dfrac{1}{2}x = 3$

4. $12x = 24$

5. $18 - x = 9$

DAILY HOMEWORK QUIZ

For use after Lesson 1.5, pages 30–35

Write the phrase as a variable expression. Let *n* represent the number.

1. 80 less than a number

2. 27 times a number

Write the sentence as an equation or inequality. Let *y* represent the number.

3. 16 more than a number is less than 35.

4. The quotient of a number and 50 is 3.

5. Tom charges $6.50 an hour to baby-sit. On Saturday, he earned $26. How many hours did he work?

NAME _____ DATE _____

Application Lesson Opener
For use with pages 36–41

Choose the equation or inequality that can be used to solve the problem. Explain why your choice is correct.

1. Nutritionists recommend drinking at least one quart of water daily. How many eight ounce glasses would that be?

A. | 8 ounces of water | ÷ | Number of glasses | = | One quart |

B. | One quart | ÷ | 8 ounces of water | = | Number of glasses |

C. | Number of glasses | · | One quart | = | 8 ounces of water |

2. A healthy diet should have no more than 30% of the calories come from fat. For example, if a cracker is 10 calories, no more than 3 calories should be fat calories. What is the most fat calories that a 150 calorie snack should have?

A. | Number of fat calories in snack | ÷ | Number of calories in snack | ≥ | Percentage of fat calories |

B. | Number of fat calories in snack | · | Percentage of fat calories | ≥ | Number of fat calories |

C. | Percentage of fat calories | · | Number of fat calories in snack | ≤ | Number of fat calories |

NAME _____ DATE _____

Practice A
For use with pages 36–41

In Exercises 1–4, which equation correctly models the situation?

1. *Bake Sale* You want to make six dozen cookies for a bake sale. If you fol-
   low the recipe, one batch makes two dozen cookies. Let b be the number of
   batches you need to bake.

   a. $2b = 6$ **b.** $\dfrac{b}{6} = 2$

2. *Height* You are 65 inches tall. You are 18 inches taller than your younger sister. Let h
   be your sister's height in inches.

   a. $h - 18 = 65$ **b.** $h + 18 = 65$

3. *Reading* Your teacher gave you two weeks to read a 252-page book. How many pages
   a night will you have to read? Let p be the number of pages read.

   a. $252 - p = 14$ **b.** $p = \dfrac{252}{14}$

4. *Car wash* The chess club is holding a car wash to raise funds for an upcoming tour-
   nament. If the club wants to raise $600, how many cars will they have to wash if they
   charge $3 for each car? Let c be the number of cars washed.

   a. $600 = \dfrac{c}{3}$ **b.** $600 = 3c$

Airplane Speed   **In Exercises 5–8, use the following information.**

A commercial airplane has been flying for two hours and has flown a
distance of 360 miles. How fast has it been flying?

Verbal Model: | Speed of airplane | • | Flight time | = | Distance traveled |

5. Assign labels to the three parts of the verbal model.

6. Use the labels to translate the verbal model into an algebraic model.

7. Use mental math to solve the equation.

8. Check to see if your answer is reasonable.

NAME _____ DATE _____

Practice B

For use with pages 36–41

In Exercises 1–5, which equation correctly models the situation?

1. *Model Planes* Your model plane collection consists of 15 models. Each plane is either a propeller plane or a jet. There are seven fewer propeller planes than jets. Let x be the number of jets.

   a. $x + (x - 7) = 15$ **b.** $x - 7 = 15$

2. *Music* An eighth note is played twice as long as a quarter note. Eight eighth notes can be played in one measure of music. Let q be the number of quarter notes played in one measure of music.

   a. $2q = 8$ **b.** $\dfrac{q}{2} = 8$

3. *Money* You have the same number of dimes as you do quarters. If you have $2.80, how many of each coin do you have? Let c be the number of coins.

   a. $\$2.80 - c = \$.10 + \$.25$ **b.** $c(\$.10 + \$.25 = \$2.80$

4. *Baseball* A baseball team's standing is based on their win-loss record. Dividing their number of wins by the total number of games they've played is how their winning percentage is calculated. If the team's percentage is 0.850, and they've played 80 games, how many games have they won? Let w be the number of games won.

   a. $\dfrac{w}{80} = 0.850$ **b.** $\dfrac{w}{0.850} = 80$

5. *Nutrition* In comparing butter to reduced-fat cream cheese, you discover that, gram for gram, butter has more than 3 times the fat of cream cheese. If the cream cheese has 6 grams of fat, what amount of fat does the butter have? Let f be the amount of fat in the butter.

   a. $f > 6 - 3$ **b.** $f > 3(6)$

Recycling In Exercises 6–8, use the following information.

A recycling center pays $.10 per aluminum can. You were paid $5.00 for recycling cans.

6. Write a verbal model that relates the amount of money paid per can, the number of cans recycled, and the total amount of money you were paid.

7. Assign labels and write an algebraic model based on your verbal model.

8. Use mental math to solve the equation. Check to see if your answer is reasonable.

NAME _____ DATE _____

Reteaching with Practice

For use with pages 36–41

GOAL Model and solve real-life problems.

> **VOCABULARY**
>
> Writing algebraic expressions, equations, or inequalities that represent real-life situations is called **modeling**. First you write a **verbal model** using words. Then you translate the verbal model into an **algebraic model**.

EXAMPLE 1 *Write an Algebraic Model*

A movie theater charges $6 admission. The total sales on a given day were $420. How many admission tickets were sold that day?

SOLUTION

Verbal Model

Cost per Ticket	•	Number of tickets	=	Total sales

Labels

Cost per ticket = 6   (dollars)

Number of tickets = n (tickets)

Total sales = 420   (dollars)

Algebraic Model

$6n = 420$ Write algebraic model.

$n = 70$ Solve using mental math.

The number of tickets sold is 70.

Exercises for Example 1

In Exercises 1 and 2, do the following.

  a. **Write a verbal model.**

  b. **Assign labels and write an algebraic model based on your verbal model.**

  c. **Use mental math to solve the equation.**

  1. The student government is selling baseball hats at $8 each. The group wants to raise $2480. How many hats does the group need to sell?

  2. You and your two sisters bought a gift for your brother. You paid $7.50 for your share (one-third of the gift). What was the total cost of the gift?

NAME _____ DATE _____

Reteaching with Practice

For use with pages 36–41

EXAMPLE 2 *Write an Algebraic Model*

You are hiking on a trail. You hike at an average speed of 1.5 miles per hour for 2.8 miles. During the last 3.4 miles, you increase your average speed by 0.2 miles per hour. How long will it take you to walk the last 3.4 miles?

SOLUTION

Verbal Model

$$\left(\boxed{\begin{array}{c} \text{Speed for} \\ \text{first 2.8 miles} \end{array}} + 0.2 \right) \cdot \boxed{\text{Time}} = \boxed{\text{Distance}}$$

Labels

Speed for first 2.8 miles = 1.5   (miles per hour)

Time = x (hours)

Distance = 3.4   (miles)

Algebraic Model

$(1.5 + 0.2) \cdot (x) = 3.4$ Write algebraic model.

$1.7x = 3.4$ Simplify.

$x = 2$ Solve using mental math.

It will take 2 hours.

Exercise for Example 2

3. A car travels at an average speed of 45 miles per hour for 8 miles, reduces its speed by 15 miles per hour for the next 4 miles, and then returns to a speed of 45 miles per hour. How long does the car travel at the reduced speed?

NAME _____ DATE _____

Quick Catch-Up for Absent Students
For use with pages 36–41

The items checked below were covered in class on (date missed) _____

Lesson 1.6: A Problem Solving Plan Using Models

_____ **Goal:** Model and solve real-life problems.

Material Covered:

     _____ Studen Help: Study Tip

     _____ Example 1: Write an Algebraic Model

     _____ A Problem Solving Plan Using Models

     _____ Student Help: Study Tip

     _____ Example 2: Write an Algebraic Model

     _____ Example 3: Write an Algebraic Model

Vocabulary:

     modeling, p. 36   algebraic model, p. 36

     verbal model, p. 36

_____ Other (specify) _____

_____

Homework and Additional Learning Support

     _____ Textbook (specify) <u>pp. 39–41</u> _____

_____

     _____ *Reteaching with Practice* worksheet (specify exercises) _____

     _____ *Personal Student Tutor* for Lesson 1.6

NAME _____ DATE _____

Real-Life Application:
When Will I Ever Use This

For use with pages 36–41

Leasing a car.

You decide to lease a car, and there are several plans from which to choose. At the end of the 4-year lease, you may purchase the car for a pre-determined amount (p), or you can turn the car back in to the dealer. With each plan you must pay a down payment (d), and a monthly payment (i) for the duration of the lease. Some plans have different mileage options (y) maximizing the average number of miles you can accumulate per year. If you exceed that amount, you are responsible for the per-mile excess charge of $.10.

Below is the table for the various leasing plans.

Option	Down Payment	Monthly Payment	Yearly Mileage Allowance	Buy-Out Purchase Price
A	$1,000.00	$419.00	12,000	$25,500
B	$2,000.00	$409.00	12,000	$24,500
C	$3,000.00	$389.00	15,000	$23,500
D	$4,000.00	$369.00	15,000	$22,500

In Exercises 1–5, use the above pricing information as well as the following information.

1. Write an algebraic model to calculate any additional mileage fees (f) given mileage (m).

2. After 4 years you decide not to purchase the car. Calculate what you owe the dealer under plans A & B if you've driven the car 53,865 miles.

3. Using the table above, write an algebraic model to calculate the total purchase price (t) of the car, assuming you do not exceed the yearly mileage allowance.

4. **a.** Using the model in Exercise 3, calculate the total purchase price for the car for each plan.

   b. Under which plan would you pay the most?

   c. Under which plan would you pay the least?

   d. Suppose the selling price of the car is $38,500. How much would you save if you were to purchase the car outright vs. Plan D? (Exclude any interest or surcharges.)

5. Suppose you chose Plan C. Calculate how many miles you can drive the car until your total price is equal to plan B.

NAME _____ DATE _____

Challenge: Skills and Applications

For use with pages 36–41

In Exercises 1–6, use the following information.

You are trying to save $20 a week to buy a new CD player. During the last 4 weeks you have saved $35, $15, $10, and $12. You want to know how much you need to save this week to average $20 for the 5 weeks.

1. How much have you saved so far?

2. Write a verbal model that relates the money you have saved so far, the money you need to save this week, the number of weeks, and the average savings that is your goal.

3. Assign labels to the parts of the verbal model.

4. Use the labels to translate the verbal model into an algebraic model.

5. Use mental math to solve the equation.

6. Interpret your solution.

In Exercises 7–9, use the following information.

Maurice's Music Store has selected CDs on sale for $9.50 each plus 5.75% sales tax. You have $48 you can spend on CDs.

7. Write an inequality that shows how many CDs you can buy.

8. Find the largest whole number that is a solution to the inequality. Use estimation and mental math to help.

9. How many CDs can you buy?

In Exercises 10–12, use the following information.

Olga Weatherby is on a road to Houston that has a speed limit of 65 miles per hour. She is 143 miles from Houston and would like to be there in two hours.

10. Can Olga make it to Houston in 2 hours if she drives at the speed limit?

11. Can Olga make it to Houston in 2.25 hours if she drives at the speed limit?

12. Write an inequality to model this situation, where t is driving time to Houston.

Lesson Plan

1-day lesson (See *Pacing the Chapter,* TE page 1A)   **For use with pages 42–47**

GOAL **Organize data using a table or graph.**

State/Local Objectives _____

_____

✓ **Check the items you wish to use for this lesson.**

STARTING OPTIONS

_____ Homework Check (1.6): TE page 39; Answer Transparencies

_____ Homework Quiz (1.6): TE page 41, CRB page 84, or Transparencies

_____ Warm-Up: CRB page 84 or Transparencies

TEACHING OPTIONS

_____ Lesson Opener: CRB page 85 or Transparencies

_____ Examples 1–3: SE pages 42–44

_____ Extra Examples: TE pages 43–44 or Transparencies

_____ Checkpoint Exercises: SE pages 42–44

_____ Concept Check: TE page 44

_____ Guided Practice Exercises: SE page 45

APPLY/HOMEWORK

Homework Assignment

_____ Transitional: SRH p. 779 1–3, 6–8; pp. 45–47, Exs. 6, 7, 11–14, 20–22, 23–43 odd

_____ Average: pp. 45–47, Exs. 8–14, 18, 20–22, 25, 29–31, 38–40

_____ Advanced: pp. 45–47, Exs. 14–22, 32–34, 41–43; EC: CRB p. 93

Reteaching the Lesson

_____ Practice Masters: CRB pages 86–87 (Level A, Level B)

_____ Reteaching with Practice: CRB pages 88–89 or Practice Workbook with Examples; Resources in Spanish

_____ Personal Student Tutor: CD-ROM

Extending the Lesson

_____ Learning Activity: CRB page 91

_____ Interdisciplinary/Real-Life Applications: CRB page 92

_____ Challenge: CRB page 93

ASSESSMENT OPTIONS

_____ Daily Quiz (1.7): TE page 47, CRB page 96, or Transparencies

_____ Standardized Test Practice: SE page 47; STP Workbook; Transparencies

Notes _____

_____

_____

Algebra 1
Chapter 1 Resource Book

TEACHER'S NAME _____ CLASS _____ ROOM _____ DATE _____

Lesson Plan for Block Scheduling
Half-block lesson (See *Pacing the Chapter,* TE page 1A)   For use with pages 42–47

GOAL   Organize data using a table or graph.

State/Local Objectives _____
_____
_____

✓ **Check the items you wish to use for this lesson.**

STARTING OPTIONS
_____ Homework Check (1.6): TE page 39; Answer Transparencies
_____ Homework Quiz (1.6): TE page 41,
        CRB page 84, or Transparencies
_____ Warm-Up: CRB page 84 or Transparencies

TEACHING OPTIONS
_____ Lesson Opener: CRB page 85 or Transparencies
_____ Examples 1–3: SE pages 42–44
_____ Extra Examples: TE pages 43–44 or Transparencies
_____ Checkpoint Exercises: SE pages 42–44
_____ Concept Check: TE page 44
_____ Guided Practice Exercises: SE page 45

APPLY/HOMEWORK
Homework Assignment (See also the assignment for Lesson 1.8.)
_____ Block Schedule: pp. 45–47 Exs. 8–14, 18, 20–22, 25, 29–31, 38–40

Reteaching the Lesson
_____ Practice Masters: CRB pages 86–87 (Level A, Level B)
_____ Reteaching with Practice: CRB pages 88–89 or Practice Workbook with
        Examples; Resources in Spanish
_____ Personal Student Tutor: CD-ROM

Extending the Lesson
_____ Learning Activity: CRB page 91
_____ Interdisciplinary/Real-Life Applications: CRB page 92
_____ Challenge: CRB page 93

ASSESSMENT OPTIONS
_____ Daily Quiz (1.7): TE page 47, CRB page 96, or Transparencies
_____ Standardized Test Practice: SE page 47; STP Workbook; Transparencies

Notes _____
_____
_____

CHAPTER PACING GUIDE	
Day	Lesson
1	1.1 (all); 1.2 (all)
2	1.3 (all); 1.4 (all)
3	1.5 (all); 1.6 (all)
4	**1.7 (all)**; 1.8 (all)
5	Ch. 1 Review and Assess

NAME _____ DATE _____

WARM-UP EXERCISES

For use before Lesson 1.7, pages 42–47

The table represents U.S. population projections for 2020–2040, in millions.

1. What is the projected population in 2020?

2. What is the projected population in 2040?

Year	Population
2020	322.7
2030	346.9
2040	370.0

DAILY HOMEWORK QUIZ

For use after Lesson 1.6, pages 36–41

Eight friends went to a restaurant for dinner. The waiter gave them a bill for $130. At the register, $30 was added to the bill for tax and tip.

1. Write a verbal model that relates the number of friends, each friend's share of the bill, the bill, and the amount added for tax and tip.

2. Assign labels to your verbal model. Let d be each friend's share.

3. Use the labels to translate your verbal model into an equation.

4. Use mental math to solve the equation.

Application Lesson Opener

For use with pages 42–47

The data in the table shows the number of endangered species of animals in the United States as of August 31, 1998.

Group	Number of species
Arachnids	5
Amphibians	9
Reptiles	14
Snails	15
Crustaceans	16
Insects	28
Mammals	59
Clams	61
Fishes	68
Birds	75

1. Which group has the least number of endangered species? Explain how you know.

2. Which group has the greatest number of endangered species? Explain how you know.

3. Suppose the names of the groups are listed in alphabetical order. Would Questions 1 and 2 be easier or harder to answer? Explain.

The bar graph shows percent of United States households subscribing to basic cable television.

4. In which year was the percent of U.S. households subscribing to basic cable television the greatest? What was the percent?

5. In which year was the percent of U.S. households subscribing to basic cable television the least? What was the percent?

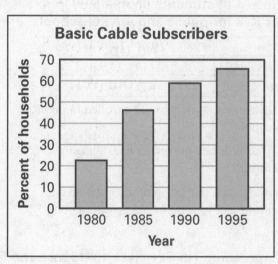

Basic Cable Subscribers

NAME _____ DATE _____

Practice A

For use with pages 42–47

Transportation   In Exercises 1–3, use the bar graph, which shows the type of transportation students used to go to school at Adams High School in a recent year.

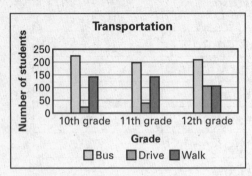

1. Which class has the most students driving to school?

2. Which two classes have the same number of students walking to school?

3. What is the most common form of transportation among 10th graders?

Ratios   In Exercises 4–6, use the bar graph, which shows the pupil-to-teacher ratio for public schools in 1985, 1990, and 1995.

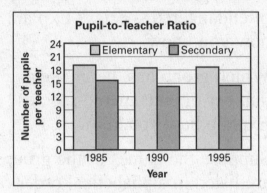

4. Which year had the highest pupil-to-teacher ratio for elementary schools?

5. What was the approximate pupil-to-teacher ratio for secondary schools in 1995?

6. Which pupil-to-teacher ratio had the largest overall decrease from 1985 through 1990?

Schools   The table gives the fall enrollment of students (in millions) in grades K–8 for public and private schools.

Year	1980	1985	1990	1995
Public school	27.6	27.0	29.9	32.4
Private school	4.0	4.2	4.1	4.4

7. How many students were enrolled in public schools in 1985?

8. What year had the fewest total students enrolled?

9. Construct a bar graph of the combined enrollment for public and private schools.

Business   The table gives a company's revenue and expenses (in thousands of dollars) for six months.

Month	July	Aug.	Sept.	Oct.	Nov.	Dec.
Revenue	48	39	36	42	57	62
Expenses	38	58	54	48	37	51

10. In what month were the expenses the greatest?

11. Profit is the difference of revenue and expenses. What was the company's profit in November?

12. Construct a line graph of the company's revenue.

Life Expectancy   In Exercises 13–15, use the line graph, which gives the life expectancy in the United States for a child (at birth).

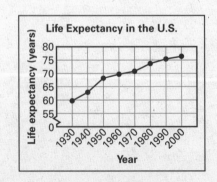

13. In which decade did life expectancy increase the least?

14. In which decade did life expectancy increase the most?

15. Discuss what the line graph shows.

NAME _____ DATE _____

Practice B

For use with pages 42–47

Transportation **In Exercises 1–3, use the bar graph, which shows the type of transportation students used to go to school at Adams High School in a recent year.**

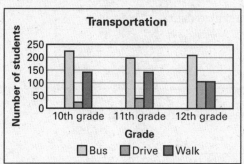

1. Which class has the most students riding the bus to school?

2. Which two classes have the same number of students walking to school?

3. What is the most common form of transportation among 12th graders?

Schools **The table gives the fall enrollment of students (in millions) in grades K–8 for public and private schools.**

Year	1980	1985	1990	1995
Public school	27.6	27.0	29.9	32.4
Private school	4.0	4.2	4.1	4.4

In Exercises 7 and 8, decide whether each statement is *true* or *false* based on the table.

7. The ratio of public to private enrollment has decreased overall from 1980 through 1995.

8. The total number of students has increased.

9. Construct a bar graph of the combined enrollment for public and private schools.

Ratios **In Exercises 4–6, use the bar graph, which shows the pupil-to-teacher ratio for public schools in 1985, 1990, and 1995.**

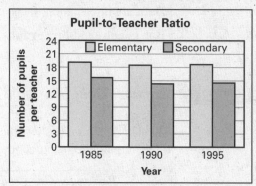

4. Which year had the highest pupil-to-teacher ratio for secondary schools?

5. What was the approximate pupil-to-teacher ratio for elementary schools in 1990?

6. Which pupil-to-teacher ratio had the largest overall decrease from 1985 through 1990?

Business **The table gives a company's revenue and expenses (in thousands of dollars) for six months.**

Month	July	Aug.	Sept.	Oct.	Nov.	Dec.
Revenue	48	39	36	42	57	62
Expenses	38	58	54	48	37	51

10. In what month were the expenses the least?

11. Profit is the difference of revenue and expenses. What was the company's profit in December?

12. Construct a line graph of the company's revenue.

13. Was it a profitable six months for the company? Explain.

Life Expectancy **In Exercises 14 and 15, use the information in the table, which gives the life expectancy in the United States for a child (at birth).**

Year	1930	1940	1950	1960	1970	1980	1990	2000
Life expectancy (years)	59.7	62.9	68.2	69.7	70.8	73.7	75.4	76.4

14. In which decade did life expectancy increase the most?

15. Construct a line graph of these data and discuss what the line graph shows.

LESSON 1.7

Reteaching with Practice

For use with pages 42–47

GOAL Organize data using a table or graph.

VOCABULARY

Data are information, facts, or numbers that describe something.

Bar graphs and **line graphs** are used to organize data.

EXAMPLE 1 *Organize Data in a Table*

The data in the table show the number of passenger cars produced by three automobile manufacturers.

Passenger Car Production (in thousands)						
Year	1970	1975	1980	1985	1990	1995
Company A	1273	903	639	1266	727	577
Company B	2017	1808	1307	1636	1377	1396
Company C	2979	3679	4065	4887	2755	2515

a. During what year was the highest total passenger car production?

b. During what year was the lowest total passenger car production?

SOLUTION

Add another row to the table. Enter the total passenger car production of all three companies.

Year	1970	1975	1980	1985	1990	1995
Total	6269	6390	6011	7789	4859	4488

a. From the table, you can see that the total passenger car production was the highest in 1985.

b. From the table, you can see that the total passenger car production was the lowest in 1995.

Exercises for Example 1

In Exercises 1 and 2, use the data from Example 1.

1. During what year was the the total passenger car production by Company B the highest?

2. During what year was the total passenger car production by Company C the lowest?

Reteaching with Practice

For use with pages 42–47

EXAMPLE 2 *Make and Interpret a Line Graph*

Use the data from Example 1. Draw a line graph to organize the data for Company A's passenger car production.

a. During which 5-year period did Company A's passenger car production decrease the least?

b. During which 5-year period did Company A's passenger car production decrease the most?

SOLUTION

Draw the vertical scale from 0 to 1400 thousand cars in increments of 200 thousand cars. Mark the number of years on the horizontal axis starting with 1970. For each number of passenger cars produced, draw a point on the graph. Then draw a line from each point to the next point.

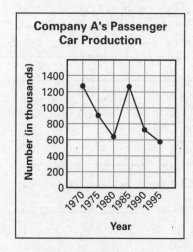

a. Company A's passenger car production decreased the least from 1990–1995.

b. Company A's passenger car production decreased the most from 1985–1990.

Exercises for Example 2

In Exercises 3 and 4, use the data from Example 1.

3. Draw a line graph to organize the data for Company C's passenger car production. During which 5-year period did Company C's passenger car production decrease the most?

4. Draw a line graph to organize the data for Company B's passenger car production. During which 5-year period did Company B's passenger car production increase the most?

Quick Catch-Up for Absent Students

For use with pages 42–47

The items checked below were covered in class on (date missed) _____

Lessons 1.7: Table and Graphs

____ **Goal:** Organize data using a table or graph.

Material Covered:

____ Student Help: Study Tip

____ Example 1: Organize Data in a Table

____ Student Help: Vocabulary Tip

____ Example 2: Interpret a Bar Graph

____ Example 3: Make and Interpret a Line Graph

____ Student Help: Study Tip

Vocabulary:

data, p. 42   bar graph, p. 43   line graph, p. 44

____ Other (specify) _____

_____

Homework and Additional Learning Support

____ Textbook (specify) pp. 45–47 _____

_____

____ Internet: Extra Examples at www.mcdougallittell.com

____ *Reteaching with Practice* worksheet (specify exercises)_____

____ *Personal Student Tutor* for Lesson 1.7

NAME _____ DATE _____

Learning Activity

For use with pages 42–47

GOAL **Make a circle graph to display the results of an in-class survey.**

Materials: compass, protractor, paper, pencil, colored pencils or markers

Exploring Circle Graphs

Another way to display information is to use a circle graph. In this activity, your group will survey your class on a specific topic and then make a circle graph to display the results of your survey.

Instructions

Select a topic that you would like to survey members of your class about.

Select categories for your topic.

Survey your class. Record the number of responses for each category and the total number of responses for the survey.

Divide the number of responses for a category by the total number of responses for the survey. Multiply the result by 360°.

Use the protractor to measure sections of the circle graph for each category using the number of degrees found in Step 4. Label the sections and give the graph an appropriate title.

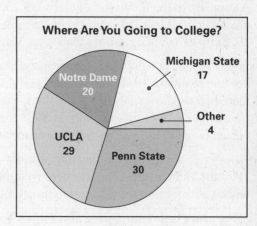

Where Are You Going to College?

Analyzing the Results

1. Why do many newspapers use circle graphs? What other forms of graphs are used in the media?

2. Discuss advantages and disadvantages to using the graphs you have learned about in this lesson, including circle graphs.

Lesson 1.7

Interdisciplinary Application

For use with pages 42–47

Space Exploration

SCIENCE On October 4, 1957, the Soviet Union launched *Sputnik I,* Earth's first artificial satellite; the Space Age had officially begun. U.S. President Dwight D. Eisenhower called for an acceleration of projects in an attempt to match Russian capabilities as soon as possible.

In January of 1958 the United States launched their first satellite, *Explorer I.* Yuri Gagarin, a Russian cosmonaut, made the first orbital flight on April 12, 1961. In May of 1961, Alan B. Shepard, Jr., became the United States' first man in space. In February of 1962, John Glenn became the first American to orbit Earth. He would later become the country's oldest astronaut when he returned to space in 1998. On July 20, 1969, *Apollo 11* brought Neil Armstrong and Edwin Aldrin, Jr., to the moon for a lunar walk. *Skylab,* the U.S. satellite lab, was placed in orbit in 1973, and a joint space mission with Russia took place in 1975. In 1976 Viking probes landed on Mars.

In the 1970's, development of the Space Shuttle system began. The system became reality in 1981 with the successful *Columbia* mission. By mid 1998 there had been over 90 shuttle missions, even with a two-year hiatus after the *Challenger*'s explosion in 1986. In 1983 the European-built *Spacelab* was placed in orbit. In 1986 the Russian space station *Mir* was established.

Voyager 2, a space probe, flew past Uranus in 1986 and Neptune in 1989. In 1990 the Hubble telescope was deployed in space. Various U.S. missions to launch new satellites or maintain present units still continue today.

Federal Outlay for Space and Other Technology in Billions of Dollars (1970–2003)											
Year	1970	1975	1980	1983	1984	1985	1986	1987	1988	1989	1990	1991
Current $	3.6	3.0	4.5	6.3	6.5	6.6	6.8	7.0	8.4	10.2	11.6	13.0

Year	1992	1993	1994	1995	1996	1997	1998	1999	2000	2001	2002	2003
Current $	12.8	13.1	12.4	12.6	12.7	13.1	12.9	12.8	12.3	12.4	12.4	12.5

* Estimated amounts for 1999–2003

1. Make a line graph using the above data on current dollars. Let each line on the horizontal axis represent one year. Let each line on the vertical axis represent 0.5 billion dollars.

2. For how many years does it appear that the general trend was to increase spending? During which 5-year period was there the greatest increase? How much did it increase?

3. What appears to be the trend from 1998 to 2003?

4. Using the information provided in the paragraph on the space program, why do you think the money spent in 1970 was so high? What might have caused the increase from 1987 to 1991?

NAME _____ DATE _____

Challenge: Skills and Applications

For use with pages 42–47

Use the information in the table.

	Joe	*Tina*	*Ali*	*Keesha*
Freshman	54	38	39	35
Sophomore	52	42	47	51
Junior	55	68	52	57
Senior	41	75	58	64

1. Find the average shooting percentage for each player.

2. Make a bar graph of the averages from Exercise 1. Make the scale from 0 to 100 by tens.

3. Make another bar graph of the averages from Exercise 1. Make the scale range from 45 to 60 by ones.

4. What impression does the graph from Exercise 2 give? What impression does the graph from Exercise 3 give?

5. Make a bar graph showing all the data in the table. Include a key which shows how the bars are shaded for each of the four years.

6. Which player did not improve?

7. Which player improved the most?

8. Who improved more, Ali or Keesha?

9. Between which two years did Tina improve the most?

10. Between which two years did Joe's percentage drop the most?

TEACHER'S NAME _____ CLASS _____ ROOM _____ DATE _____

Lesson Plan

1-day lesson (See *Pacing the Chapter,* TE page 1A)   **For use with pages 48–54**

GOAL **Use four different ways to represent functions.**

State/Local Objectives _____

_____

✓ **Check the items you wish to use for this lesson.**

STARTING OPTIONS
_____ Homework Check (1.7): TE page 45; Answer Transparencies
_____ Homework Quiz (1.7): TE page 47, CRB page 96, or Transparencies
_____ Warm-Up: CRB page 96 or Transparencies

TEACHING OPTIONS
_____ Lesson Opener: CRB page 97 or Transparencies
_____ Examples 1–3: SE pages 48–50
_____ Extra Examples: TE pages 49–50 or Transparencies; Internet Help at www.mcdougallittell.com
_____ Checkpoint Exercises: SE page 50
_____ Concept Check: TE page 50
_____ Guided Practice Exercises: SE page 51

APPLY/HOMEWORK
Homework Assignment
_____ Transitional: pp. 51–54, Exs. 7, 8, 13, 16, 17, 20–22, 27–32, 36–40, Quiz 3
_____ Average: pp. 51–54, Exs. 9, 10, 14, 15, 18, 23, 27–32, 36–40, Quiz 3
_____ Advanced: pp. 51–54, Exs. 11, 12, 15, 19, 24–29*, 33–35, 44–46, Quiz 3; EC: CRB p. 105

Reteaching the Lesson
_____ Practice Masters: CRB pages 98–99 (Level A, Level B)
_____ Reteaching with Practice: CRB pages 100–101 or Practice Workbook with Examples; Resources in Spanish
_____ Personal Student Tutor: CD-ROM

Extending the Lesson
_____ Learning Activity: CRB page 103
_____ Interdisciplinary/Real-Life Applications: CRB page 104
_____ Challenge: CRB page 105

ASSESSMENT OPTIONS
_____ Daily Quiz (1.8): TE page 53 or Transparencies
_____ Standardized Test Practice: SE page 53; STP Workbook; Transparencies
_____ Quiz 1.6–1.8: SE page 54

Notes _____

_____

_____

Algebra 1
Chapter 1 Resource Book

TEACHER'S NAME _____ CLASS _____ ROOM _____ DATE _____

Lesson Plan for Block Scheduling

Half-block lesson (See *Pacing the Chapter,* TE page 1A)   **For use with pages 48–54**

GOAL   **Use four different ways to represent functions.**

State/Local Objectives _____

_____

_____

✓ **Check the items you wish to use for this lesson.**

STARTING OPTIONS

____ Homework Check (1.7): TE page 45; Answer Transparencies

____ Homework Quiz (1.7): TE page 47,
     CRB page 96, or Transparencies

____ Warm-Up: CRB page 96 or Transparencies

TEACHING OPTIONS

____ Lesson Opener: CRB page 97 or Transparencies

____ Examples 1–3: SE pages 48–50

____ Extra Examples: TE pages 49–50 or Transparencies; Internet Help
     at www.mcdougallittell.com

____ Checkpoint Exercises: SE page 50

____ Concept Check: TE page 50

____ Guided Practice Exercises: SE page 51

APPLY/HOMEWORK

Homework Assignment (See also the assignment for Lesson 1.7.)

____ Block Schedule: pp. 51–54, Exs. 9, 10, 14, 15, 18, 23, 27–32, 36–40, Quiz 3

Reteaching the Lesson

____ Practice Masters: CRB pages 98–99 (Level A, Level B)

____ Reteaching with Practice: CRB pages 100–101 or Practice Workbook
     with Examples; Resources in Spanish

____ Personal Student Tutor: CD-ROM

Extending the Lesson

____ Learning Activity: CRB page 103

____ Interdisciplinary/Real-Life Applications: CRB page 104

____ Challenge: CRB page 105

ASSESSMENT OPTIONS

____ Daily Quiz (1.8): TE page 53 or Transparencies

____ Standardized Test Practice: SE page 53; STP Workbook; Transparencies

____ Quiz 1.6–1.8: SE page 54

Notes _____

_____

_____

CHAPTER PACING GUIDE	
Day	Lesson
1	1.1 (all); 1.2 (all)
2	1.3 (all); 1.4 (all)
3	1.5 (all); 1.6 (all)
4	1.7 (all); **1.8 (all)**
5	Ch. 1 Review and Assess

NAME _____ DATE _____

WARM-UP EXERCISES

For use before Lesson 1.8, pages 48–54

Evaluate the expression for the given value of the variable.

1. $12 + 3x$ when $x = 0$

2. $12 + 3x$ when $x = 1$

3. $12 + 3x$ when $x = 2$

4. $12 + 3x$ when $x = 3$

5. Describe a pattern in the answers for Exs. 1–4.

DAILY HOMEWORK QUIZ

For use after Lesson 1.7, pages 42–47

The data in the table represents the number of households in the U.S. in millions.

Year	'60	'70	'80	'90	'00
No.	52.8	63.4	80.8	93.3	103.2

a. Draw a bar graph.

b. During which 10 year period did the number of households increase the most?

Lesson 1.8

Visual Approach Lesson Opener

For use with pages 48–54

SET UP: Work in a group.
YOU WILL NEED: • **masking tape or string** • **index cards**

Create an input-output function machine
with members of your group. Place masking
tape or string on the floor to make a table
like the one show at the right. Use index
cards as labels.

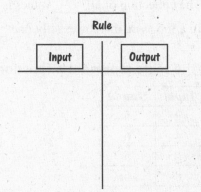

An addition function rule can be shown by
students standing on the vertical tape. For
example, two students standing on the tape
represents the rule "add two students." The
input is shown by the number of students
standing in the input column. To make the
function machine work, the "input" joins
the "rule" in the output column.

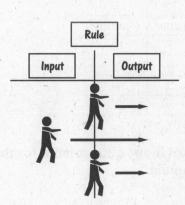

1. Show the rule "add two students." Show an input of 1. Use
   your machine to find the output. Repeat for inputs of 2, 3,
   and 4.

2. Use your function machine to find the output for the rule
   "add one student" for inputs of 1, 2, 3, and 4.

3. Use your function machine to find the output for the rule
   "add three students" for inputs of 1, 2, 3, and 4.

NAME _____ DATE _____

Practice A

For use with pages 48–54

Complete the sentence.

1. A __?__ is a relationship between two quantities, called the input and output.

2. The collection of all __?__ values is called the domain of the function.

3. The collection of all __?__ values is called the range of the function.

4. In a function, there is exactly one __?__ for each __?__.

Does the table represent a function? Explain.

5.
Input	Output
2	5
4	6
6	5
8	6

6.
Input	Output
5	0
4	0
3	0
2	0

7.
Input	Output
1	1
1	2
2	3
2	4

8.
Input	Output
3	6
4	5
5	4
6	3

9.
Input	Output
4	3
13	7
9	4
4	0

10.
Input	Output
1	1
0	0
5	5
2	2

Make an input-output table for the function. Use 0, 1, 2, and 3 as the domain.

11. $y = 2x + 3$

12. $y = 3x$

13. $y = 9 - x$

14. $y = x + 5$

15. $y = x + \frac{1}{2}$

16. $y = 10 - 3x$

Aerobics Class **In Exercises 17–19, use the following information.**

You join an aerobics class at the local gym. The cost is $3 per class plus $10 for the initial membership fee.

17. Write an equation that shows the relationship between the number of classes n you attend and the amount you pay p.

18. Evaluate the equation for $n = 1, 2, 5, 8,$ and 10. Organize your results in an input-output table.

19. Draw a line graph to represent the data in the input-output table.

Monarch Butterflies **In Exercises 20 and 21, use the following information.**

When the monarch butterfly is migrating to the south, it has an average speed of 80 miles per day.

20. Write an equation that shows the relationship between the number of days t and the distance (in miles) it has traveled d.

21. Evaluate the equation for $t = 2, 5, 8,$ and 10. Organize your results in an input-output table.

NAME _____ DATE _____

Practice B

For use with pages 48–54

Complete the sentence.

1. A ___?___ is a relationship between two quantities, called the input and output.

2. The collection of all ___?___ values is called the domain of the function. The collection of all ___?___ values is called the range of the function.

3. In a function, there is exactly one ___?___ for each ___?___ .

4. An ___?___ is a table that lists the outputs for several different inputs.

Does the table represent a function? Explain.

5.
Input	Output
2	10
4	12
6	10
8	12

6.
Input	Output
9	0
8	0
7	0
6	0

7.
Input	Output
1	1
2	2
2	3
3	4

Make an input-output table for the function. Use 0, 1, 2, and 3 as the domain.

8. $y = 2x + 5$

9. $y = 4x$

10. $y = 15 - x$

11. $y = x + 18$

12. $y = 2x + \frac{1}{2}$

13. $y = 20 - 3x$

Make an input-output table for the function. Use 2, 2.5, 4, 5, and 5.5 as the domain.

14. $y = 3x + 1.5$

15. $y = \dfrac{22}{x} + 7$

16. $y = x^2 - 2.5$

Aerobics Class **In Exercises 17–20, use the following information.**

You join an aerobics class at the local gym. The cost is $3 per class plus $10 for the initial membership fee.

17. Write an equation that shows the relationship between the number of classes n you attend and the amount you pay p.

18. Evaluate the equation for $n = 1, 2, 5, 8,$ and 10. Organize your results in an input-output table.

19. Draw a line graph to represent the data in the input-output table.

20. Describe the domain and range of the function.

Monarch Butterflies **In Exercises 21–23, use the following information.**

When the monarch butterfly is migrating to the south, it has an average speed of 80 miles per day.

21. Write an equation that shows the relationship between the number of days t and the distance (in miles) it has traveled d.

22. Evaluate the equation for $t = 2, 5, 8,$ and 10. Organize your results in an input-output table.

23. Draw a line graph to represent the data in the input-output table.

NAME _____ DATE _____

Reteaching with Practice

For use with pages 48–54

GOAL Use four different ways to represent functions.

VOCABULARY

A **function** is a rule that establishes a relationship between two quantities, called the **input** and the **output.**

Making an **input-output table** is one way to describe a function.

The collection of all input values is the **domain** of the function.

The collection of all output values is the **range** of the function.

EXAMPLE 1 *Identify a Function*

Does the table represent a function? Explain.

Input	Output
5	1
5	2
10	3
15	4

SOLUTION

The table does not represent a function. For the input value 5, there are two output values, not one.

Exercises for Example 1

In Exercises 1 and 2, does the table represent a function? Explain.

1.

Input	Output
1	4
2	8
3	12
4	16

2.

Input	Output
1	5
2	6
2	7
3	8

Reteaching with Practice

For use with pages 48–54

EXAMPLE 2 *Make an Input-Output Table*

Make an input-output table for the function $y = 3x + 1.5$. Use 0, 1, 2, and 3 as the domain.

SOLUTION

List an output for each of the inputs.

INPUT	FUNCTION	OUTPUT
$x = 0$	$y = 3(0) + 1.5$	$y = 1.5$
$x = 1$	$y = 3(1) + 1.5$	$y = 4.5$
$x = 2$	$y = 3(2) + 1.5$	$y = 7.5$
$x = 3$	$y = 3(3) + 1.5$	$y = 10.5$

Make an input-output table.

Input x	Output y
0	1.5
1	4.5
2	7.5
3	10.5

Exercises for Example 2

In Exercises 3–5, make an input-output table for the function. Use 0, 1, 2, and 3 as the domain.

3. $y = 5 - x$ **4.** $y = 4x + 1$ **5.** $y = 9 - x$

EXAMPLE 3 *Write an Equation*

The county fair charges $4 per vehicle and $1.50 for each person in the vehicle. Represent the total charge C as a function of the number of persons p. Write an equation for the function.

SOLUTION

Verbal Model

Total charge	=	Vehicle charge	+	Rate per person	·	Number of persons

Labels
Total charge = C (dollars)
Vehicle charge = 4 (dollars)
Rate per person = 1.50 (dollars)
Number of persons = p (persons)

Algebraic Model $C = 4 + 1.5p$

Exercises for Example 3

6. Rework Example 3 if the vehicle charge is $1.50 and $4 is charged for each person in the vehicle.

7. Rework Example 3 if the vehicle charge is $3 and $2.50 is charged for each person in the vehicle.

NAME _____ DATE _____

Quick Catch-Up for Absent Students

For use with pages 48–54

The items checked below were covered in class on (date missed) _____

Lesson 1.8: An Introduction to Functions

_____ **Goal:** Use four different ways to represent functions.

Material Covered:

_____ Example 1: Make an Input-Output Table

_____ Student Help: Study Tip

_____ Example 2: Use a Table to Graph a Function

_____ Example 3: Write an Equation to Represent a Function

Vocabulary:

function, p. 48   input, p. 48

output, p. 48   input-output table, p. 48

domain, p. 49   range, p. 49

_____ Other (specify) _____

_____

Homework and Additional Learning Support

_____ Textbook (specify) _pp. 51–54_____

_____

_____ Internet: Extra Examples at www.mcdougallittel.com _____

_____ *Reteaching with Practice* worksheet (specify exercises)_____

_____ *Personal Student Tutor* for Lesson 1.8

NAME _____ DATE _____

Learning Activity

For use with pages 48–54

GOAL **To compare the circumferences and areas of circles**

Materials: One piece of coordinate graph paper, ruler, loose-leaf paper, pencil

Exploring Circumference and Area

The circumference of a circle is the distance around the circle. Circumference is measured in units such as centimeters or inches. The area of a circle can be thought of as the number of square units contained in its interior. Area is measured in square units such as square centimeters or square inches. In this activity, you will compare the circumferences and areas of several circles.

Instructions

Make a table on a piece of loose-leaf paper with the headings Radius, Circumference, and Area.

Measure the radius of the circle at the right in centimeters. Find its circumference by using the function $C = 2\pi r$. Find its area by using the function $A = \pi r^2$. Record all information in the table ($\pi \approx 3.14$).

Increase the radius of the circle by one centimeter. Find the circumference and area of the new circle. Record all information.

Increase the radius of the circle four more times, one centimeter at a time. Each time, calculate the circumference and area and record the information on the table.

Graph the two functions on coordinate graph paper.

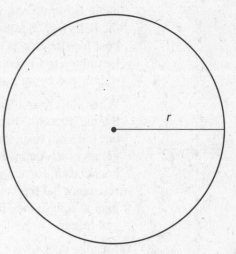

Radius	*Circumference*	*Area*

Analyzing the Results

1. What happens to the circumference of a circle as the radius increases one unit?

2. What happens to the area of a circle as the radius increases one unit?

NAME _____ DATE _____

Interdisciplinary Application

For use with pages 48–54

Paul Revere

HISTORY Paul Revere, made famous by Henry Wadsworth Longfellow's poem, "Paul Revere's Ride," was a major contributor to the Revolutionary War. The famous ride took place on April 18, 1775. England's best troops were waiting at the foot of Boston Common to take a boat trip across Back Bay. British Governor General Gage was hoping to seize the gunpowder that the patriots were storing in Concord and keep the patriots from getting support from the other New England provinces. Dr. Joseph Warren sent both Revere and William Dawes from Boston to Lexington to spread the message and warn the assumed targets, John Hancock and Samuel Adams, who were preparing to depart for the meeting of the Second Continental Congress.

On the way to Lexington, Dawes had been able to talk his way past the guards along Boston Neck, while Revere stayed downstream from the sailors guarding the Ferryway. Because of some miscommunication by the British after their boat trip, the troops were not reassembled and ready to march through Lexington to Concord until around 2:00 A.M., by which time the Lexington militia was being assembled.

When Revere and Dawes resumed their ride to Concord, they (along with Dr. Samuel Prescott) were stopped by a patrol of British officers. Dawes escaped, but fell from his horse and had to walk back to Lexington. Prescott was able to get away to continue the warning, and Revere was held at gunpoint and his horse taken. Revere was able to deceive the officers with tales of great patriot resistance, so fearing their own lives they allowed him to leave. General Gage's plan to squash the planned rebellion backfired; the American Revolutionary War had begun.

Imagine that Revere had to travel 1.5 miles to Charlestown. Assume that from Charlestown to Lexington he was able to ride a horse that traveled at a rate of 1248 feet per minute for 55 minutes. (It actually took Revere about 65 minutes to travel the distance, because he had to stop in towns to speak with local leaders.) Represent his total distance traveled (D) as a function of the time (t) in minutes that Revere traveled by horse. (*Hint:* Be sure to include the initial 1.5 miles in the equation and to convert to feet (1 mile = 5280 feet).)

1. Write an equation for the function, applying the problem-solving strategies you learned in Lesson 1.5. (Write a verbal model and assign labels, first.)

2. Use your equation to make an input-output table for the function. For inputs, begin at 35 minutes, and use the time increment of 5 minutes until you reach 55 minutes.

3. About how many total miles did Revere travel to reach Lexington that night?

4. Imagine that Revere had to walk the entire distance that night. If he could walk at a rate of 440 feet per minute, how long would it take him to reach Lexington? The distance Revere traveled D is modeled by $D = 440t$, where t represents the time in minutes. Make an input-output table for the function. For inputs, use time increments of 20 minutes. (*Hint:* Use your answer from Exercise 3 converted to feet to find Revere's total distance traveled.)

NAME _____ DATE _____

Challenge: Skills and Applications

For use with pages 46–54

For Exercises 1–4, make an input-output table for the function. Use 1, 1.25, 2.35, 2.8, and 3 as the domain.

1. $y = 7x - 2.1$

2. $y = \dfrac{x}{5} + 9.2$

3. $y = x^2 - 0.7$

4. $y = \dfrac{(x - 0.4)^2}{2}$

For Exercises 5–7, decide whether y is a function of x in the situation. Explain.

5. The amount of sales tax y on a purchase of x dollars in a specific location

6. The temperature y after a change of $10°$ from temperature x

7. The number of days y in a year x

8. Describe the range of the function in Exercise 7.

For Exercise 9 and 10, look for a pattern and write an equation for the function defined by the input-output table.

9.

Input	Output
2	7
0	3
1	4
5	28

10.

Input	Output
2	6
3	4
4	3
6	2

Lesson 1.8

Chapter Review Games and Activities

For use after Chapter 1

Crossword puzzle: Using the clues at the bottom of the page, review vocabulary from Chapter 1.

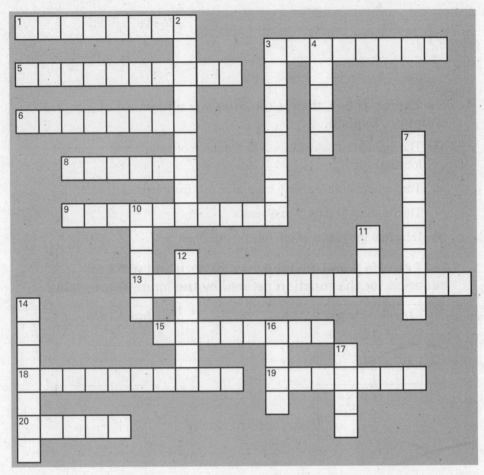

Across

1. The answer to a division problem
3. The 3 in 4^3
5. When a symbol such as $<$ or $>$ is placed between two expressions
6. A number that when substituted in an equation makes a true statement
8. The first step in modeling is to write this type of model
9. Numbers and/or variables combined with mathematical symbols
13. Parentheses or brackets
15. Letter used to represent one or more numbers
18. The answer to a subtraction problem
19. A number raised to the second power

20. A number raised to the third power

Down

2. Figures whose areas can be found using the formula $A = \frac{1}{2}bh$
3. A statement in which two expressions are equal
4. An expression of the form 5^3
7. A rule that establishes a relationship between input and output
10. Collection of all output values
11. The answer to an addition problem
12. Collection of all input values
14. The answer to a multiplication problem
16. The 6 in 6^2
17. Information, facts, or numbers that describe something

NAME _____ DATE _____

Chapter Test A

For use after Chapter 1

Evaluate the expression for the given value of the variable.

1. $q - 10$ when $q = 13$ **2.** $34x$ when $x = 4$

3. If you are drive at an average speed of 60 miles per hour for 7 hours, how far do you travel?

4. The perimeter of a rectangle is the sum of its four sides. Find the perimeter of the rectangle below.

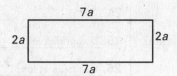

Evaluate the power.

5. 2^6 **6.** 5^3

Evaluate the expression for the given value of the variable.

7. c^5 when $c = 3$ **8.** $(2y)^3$ when $y = 3$

Evaluate the expression for the given values of the variables.

9. $x^2 + y$ when $x = 4$ and $y = 3$ **10.** $(a + b)^3$ when $a = 3$ and $b = 5$

11. A box of rice is 19 centimeters long, 12 centimeters wide, and 3 centimeters deep. The volume of the box is the product of its length, width, and height. What is the volume of the box?

Evaluate the expression.

12. $8 - 6 + 4$ **13.** $16 \div (4 - 2) - 3$

Check whether the given number is a solution of the equation or inequality.

14. $3x + 5 = 17$; 2 **15.** $4y - 7 = 3y - 4$; 3

16. $2m - 3 < 4$; 2 **17.** $5 + 2n \geq 12$; 0

Write the verbal phrase as an algebraic expression. Use x for the variable in your expression.

18. Sum of 8 and a number   **19.** Ten less than a number

Answers

1. _____
2. _____
3. _____
4. _____
5. _____
6. _____
7. _____
8. _____
9. _____
10. _____
11. _____
12. _____
13. _____
14. _____
15. _____
16. _____
17. _____
18. _____
19. _____

Review and Assess

NAME _____ DATE _____

Chapter Test A

For use after Chapter 1

Write the verbal sentence as an equation or an inequality.

20. The sum of three and x is ten.

21. Four is greater than six times a number t.

Write an equation or inequality to model the situation.

22. The distance d to school is $1\frac{1}{2}$ miles more than the distance p to the park.

23. The perimeter of a square with a side length s is greater than or equal to seventy plus five.

24. Kyra's age is at least 3 years more than Al's age.

25. The table shows the protein content of some foods. Make a bar graph of the data.

Food ($\frac{1}{2}$ *cup*)	spaghetti	rice	peanut butter
Protein (*grams*)	2	5	8

Answers

20. _____

21. _____

22. _____

23. _____

24. _____

25. Use grid at left.

26. Use table at left.

27. Use table at left.

Make an input-output table for the function. Use 0, 1, 2, and 3 as the domain.

26. $y = 7x + 2$

Input	Output

27. $y = 12 - 3x$

Input	Output

Chapter Test B

For use after Chapter 1

Evaluate the expression for the given value of the variable.

1. $25q$ when $q = 6$

2. $\dfrac{16}{x}$ when $x = 2$

3. If you drive at an average speed of 65 miles per hour for 4 hours, how far do you travel?

4. The area of a rectangle is the product of its length and width. Find the area of the rectangle below.

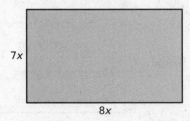

7x

8x

Evaluate the expression for the given value of the variable.

5. n^3 when $n = 4$

6. $(2x)^4$ when $x = 2$

7. $8 + 5a^2$ when $a = 7$

8. $64 - \dfrac{32}{b}$ when $b = 4$

Evaluate the expression for the given values of the variables.

9. $x + y^2$ when $x = 5$ and $y = 9$

10. $(a - b)^4$ when $a = 10$ and $b = 6$

11. A storage container is 22 inches long, 22 inches wide, and 20 inches deep. The volume of the container is the product of its length, width, and height. What is the volume of the container?

Evaluate the expression.

12. $49 \div 7 + 3 \cdot 6$

13. $4[(29 - 12) + 10]$

14. $[44 \div (10 - 8)^2] + 7$

15. $\frac{1}{2} \cdot 18 - 3^2$

Check whether the given number is a solution of the equation or inequality.

16. $16x + 3 = 29 - 3x$; 2

17. $10x - 4 \le 20$; 5

Answers

1. _____
2. _____
3. _____
4. _____
5. _____
6. _____
7. _____
8. _____
9. _____
10. _____
11. _____
12. _____
13. _____
14. _____
15. _____
16. _____
17. _____

Review and Assess

Algebra 1
Chapter 1 Resource Book

NAME _____ DATE _____

Chapter Test B

For use after Chapter 1

Write the verbal phrase as an algebraic expression. Use *x* for the variable in your expression.

18. A number increased by $\frac{1}{2}$ **19.** A number multiplied by $\frac{2}{3}$

Write the verbal sentence as an equation or an inequality.

20. Three is less than or equal to four minus a number.

21. The third power of two is eight.

Write an equation or inequality to model the situation.

22. Four hundred dollars is less than or equal to the product of $32 and the number *p* of passes to an amusement park.

23. Write an algebraic model that would be used to calculate how many minutes *m*, a class period is if there are 8 periods in a day, 5 minutes between periods, and 6 hours in a school day.

24. The table shows the price of gasoline per gallon for a four-month period. Make a line graph of the data.

Month	March	April	May	June
Price of gas (per gallon)	$1.29	$1.09	$1.19	$1.15

Answers

18. _____

19. _____

20. _____

21. _____

22. _____

23. _____

24. <u>Use grid at left.</u>

25. <u>Use table at left.</u>

26. <u>Use table at left.</u>

Make an input-output table for the function. Use 1, 1.5, 2, and 3 as the domain.

25. $y = 2x^2 + 3.1$

Input	Output

26. $y = \dfrac{6}{x} - 0.5$

Input	Output

Review and Assess

1. You hike at a rate of 3 miles per hour. Find the distance you travel in 6 hours.

   (A) 2 miles   (B) 3 miles
   (C) 18 miles   (D) 9 miles

2. What is the area of the figure?

   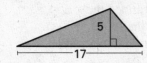

   (A) 42.5   (B) 85
   (C) 4.25   (D) 8.5

3. What is the value of $(4x)^2 + 3$ when $x = 3$?

   (A) 39   (B) 147
   (C) 153   (D) 225

4. What is the value of $\dfrac{10^2 + 8 \cdot 7}{3(14 \cdot 2 - 15)}$?

   (A) $\dfrac{52}{23}$ (B) 4

   (C) $\dfrac{187}{39}$ (D) $\dfrac{826}{39}$

5. Which of the following numbers is a solution of the equation $4x - 13 = 22 - 3x$?

   (A) $\dfrac{7}{9}$ (B) $\dfrac{9}{7}$

   (C) 5   (D) 9

6. Which of the following numbers is a solution of the inequality $6x - 7 < 43 - x$?

   (A) 7   (B) 8
   (C) 9   (D) 10

In Questions 7 and 8, choose the statement below that is true about the given numbers.

A The number in column A is greater.
B The number in column B is greater.
C The two numbers are equal.
D The relationship cannot be determined from the given information.

7.
Column A	Column B
$45 \div (9 - 4) + 3$	$45 \div 9 - 4 + 3$

   (A)   (B)   (C)   (D)

8.
Column A	Column B
3^4	$3^3 \cdot 3^1$

   (A)   (B)   (C)   (D)

9. The number of students in a history class is three less than twice the number of students in a computer class. If the number of students in the computer class is c, how many students are in the history class?

   (A) $2c$ (B) $2c - 3$
   (C) $2c + 3$ (D) $3 - 2c$

10. Which of the following represents a function?

   I.
Input	Output
2	5
3	5
4	5
5	5

   II.
Input	Output
2	4
3	6
3	8
5	10

   III.
Input	Output
1	1
2	2
3	2
4	1

   (A) All   (B) I and II
   (C) I and III   (D) II and III

Review and Assess

Alternative Assessment and Math Journal

For use after Chapter 1

JOURNAL   1. Sue and Amit compared their algebra homework and disagreed on the value of $6 + 3^2 \div 3 - 5$. Sue said the answer was 22, but Amit claimed the answer was 4. (a) Explain which answer is correct and why. (b) State the correct order of operations. Why is it important to have an established order of operations? (c) The student with the incorrect answer claimed that this answer was found using a calculator. Was the student's calculator programmed to use the established order of operations? If yes, explain why that answer was incorrect. If no, explain how the calculator determined the answer that turned out to be incorrect. Specify what the student should do in the future to determine the correct solution.

MULTI-STEP PROBLEM   2. *i.* There is a $1 entry fee to a fair. The cost of each ride ticket is $3. Sally spends exactly $10. Write an equation to represent the number of tickets Sally purchased. Solve the equation using mental math.

   ii. There is a $1 entry fee to the fair. The cost of each ride ticket is $3. Naveen spends $10 or less. Write an *inequality* to represent the possible number of tickets Naveen purchased. Solve the inequality using mental math.

   iii. There is a $1 entry fee to the fair. The cost of each ride ticket is $3. David spends less than $10. Write an *inequality* to represent the possible number of tickets David purchased. Solve the inequality using mental math.

   a. Interpret the solutions for problems *i, ii,* and *iii* in your own words.

   b. Determine the greatest number of rides for each individual.

3. *Critical Thinking* Consider the following.

   a. How do the number of rides for Naveen and David differ? What key words vary the meaning in the problems for Naveen and David?

   b. Did Sally and Naveen necessarily take the same number of rides? Explain your reasoning.

4. *Writing* Create a similar scenario for Rita's possible number of tickets that is different from the other three.

Review and Assess

JOURNAL SOLUTION

1. a–c. Complete answers should include these points.

a. $6 + 3^2 \div 3 - 5 = 6 + 9 \div 3 - 5$ Evaluate power.

$\qquad\qquad\qquad = 6 + 3 - 5$ Complete the division.

$\qquad\qquad\qquad = 9 - 5$ Add and subtract from left to right.

$\qquad\qquad\qquad = 4$ Simplify.

Amit is correct, since he used the order of operations.

b. • Explain order of operations as: First do operations that occur within grouping symbols. Then evaluate powers. Multiply and divide from left to right. Finally, add and subtract from left to right.

• Explain that it is important to have an established order of operations for there to be consistency within mathematics.

c. • Explain that the calculator was not programmed with the established order of operations and completed all calculations from left to right.

• Explain that Sue should be sure to enter each computation separately using the order of operations herself, using parentheses as needed.

MULTI-STEP PROBLEM SOLUTION

2. a. $1 + 3x = 10$; $x = 3$, Sally will purchase exactly 3 tickets.

$1 + 3x \leq 10$; $x \leq 3$, Naveen will purchase 3 tickets or less.

$1 + 3x < 10$; $x < 3$, David will purchase less than 3 tickets.

b. 3 rides for Sally and Naveen, 2 rides for David

3. a. The number of rides for both can be less than 3, but for Naveen it can also be 3. The key words are *less than $10* and *$10 or less,* they change the meaning of the inequality.

b. No; Naveen may take 3 or fewer rides; Sally must take 3.

4. *Sample answer:* Rita can spend $10 or more.

MULTI-STEP PROBLEM RUBRIC

4 Students answer all parts of the problem correctly. Clear interpretations are given for the solutions. Students are able to distinguish between an equation and an inequality. They are able to correctly compare the number of solutions. Students interpret $<$ versus $\leq$.

3 Students are able to correctly set up the models for 1, 2, and 3, although some minor errors in solutions may occur. Graphs are accurate. Students are able to interpret models. However, explanations of reasoning may be weak or minor misunderstandings may occur.

2 Students partially set up and solve the models. Some explanations and interpretations are attempted but may not be completely accurate. Differences between equations and inequalities are not explained.

1 Students are unable to set up or solve the models. Little or no explanations are given. They do not seem to understand the questions.

Review and Assess

Project: Watch It Disappear

For use after Chapter 1

OBJECTIVE   Analyze the evaporation rate of a cup of water.

MATERIALS   measuring cup, paper, pencil, graphing calculator or computer (optional)

INVESTIGATION   Measure one cup of water in the measuring cup and set it in a sunny place. Measure the water level at the same time each day for 5 days, to the nearest eighth of a cup. If your cup only shows fourths, estimate. Leave your water in the sunny place to check your predictions.

1. Complete a table like the following for your data.

Day	1	2	3	4	5
Water Level (cups)	1				
Change from previous day	___				

2. Make a line graph of your data. What conclusions can you draw from your graph?

3. Is water level a function of time in a situation like this? Explain.

4. Using the information in the row entitled "Change from previous day" in the table from Exercise 1, find the average of the changes in water level between consecutive days.

5. Write a verbal model that relates the variables. Use the average that you found in step 4 as the amount of change between any two consecutive days. Use labels to translate the verbal model into an algebraic model.

6. Use the model to predict the water level on the next school day after the fifth day. Check your prediction.

7. Write an equation you can solve to find what day the water should be completely evaporated. Use mental math to solve the equation.

PRESENT YOUR RESULTS   Write a report about your experiment. Explain how you derived your function and used it to predict when the water would be completely evaporated. Also discuss the accuracy of your predictions. Include your table and your graph.

Project: Watch It Disappear

For use after Chapter 1

GOALS
- Use tables and graphs to organize data.
- Identify a function.
- Write a verbal model and an algebraic model to solve a real-life problem.
- Evaluate a variable expression.

MANAGING THE PROJECT Ask students to bring measuring cups from home. You will need one cup per student, pair, or group. At the end of the project, you may want to compare results found with plastic cups to those found with glass ones. It is essential that cups be placed in the sun and left undisturbed. Some students may need help estimating to the nearest eighth of a cup.

Important points to address include: the table is used to find data points to plot and to look at the change in water level between days, the linear function is used as a model of the actual data points which probably were not collinear.

RUBRIC The following rubric can be used to assess student work.

4 The student makes a table that shows day number, water level, and change from previous day, then draws the line graph containing the points listed, derives an algebraic model, and uses the model to make predictions. The report shows an understanding that the relationship between day and water level is a function. All work is complete and correct.

3 The student draws the line graph, derives an algebraic model, and uses the model to make predictions. However, the report does not give a clear explanation of the experiment. There may be some minor misunderstanding of content, such as saying that there is no function because the change is not constant.

2 The student partially achieves the mathematical and project goals of drawing the line graph, deriving an algebraic model, and using the model to make predictions. However, the report indicates a limited grasp of the main ideas or requirements. There may be computation errors in finding the average. Some of the work is incomplete, misdirected, or unclear.

1 The student is unable to complete the experiment or is unable to record the data in a table. There is no understanding of the change from day to day. No verbal or algebraic model is written.

Cumulative Review

For use with Chapter 1

Evaluate the expression when $x = 3$. (1.1)

1. $x + 7$

2. $\dfrac{36}{x}$

3. $5x$

4. $x - 2$

5. $10x$

6. $12 \div x$

Evaluate the expression when $b = 5$. (1.2)

7. b^2

8. $(15 - b)^3$

9. $(7b)^2$

10. $b^4 - 600$

11. $b + b$

12. $(3b)^2$

Evaluate the expression. (1.3)

13. $3 + 20 \div 5 - 2^2$

14. $(28 \div 7)^2 - 3^2$

15. $\dfrac{10 - 3^2}{20 - 3 \cdot 4}$

16. $4^2 + 24 \div 24 - 1$

Check whether the given number is a solution of the equation or inequality. (1.4)

17. $4a - 5 = 7; 3$

18. $x^3 - x^2 = 4; 2$

19. $5y - 4 > 18; 4$

20. $2x - 5 \le 13; 6$

Write the verbal phrase or sentence as a variable expression or inequality. (1.5)

21. nine times a number n

22. x is greater than twenty.

23. quotient of a and five

24. The product of a and b is less than ten.

Solve. (1.6)

25. If you save $25 a month, how many months must you save to buy a stereo costing $225?

26. You are given $90 to buy CDs for the student dance. Each CD costs $15. How many CDs can you buy?

27. A 150-pound student burns 5.4 calories per minute jogging. If the student jogs for 30 minutes, how many calories does the student burn?

Describe the domain and range of the function. (1.8)

28.

Input	0	2	4	6	8	10
Output	7	10	12	15	17	21

29.

Input	0	10	20	30	40	50
Output	21	25	28	32	35	43

Cumulative Review

For use with Chapter 1

Use the tables below to construct a bar graph. (1.7)

30. The table below shows the number of births at a zoo for different years. Make a bar graph of the data.

Year	1994	1995	1996	1997	1998
Number of Births	8	7	12	14	10

31. The table below shows the enrollment for tenth graders at West Park High School for different years. Make a bar graph of the data.

Year	1993	1994	1995	1996	1997	1998	1999
Enrollment	50	42	31	38	45	55	60

In Exercises 32–35, use the line graph, which shows the temperature over a nine-hour period. (1.7)

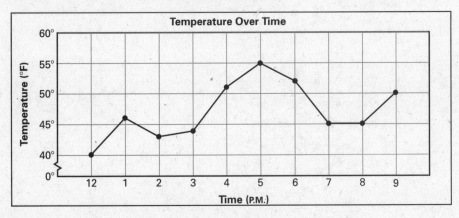

32. For how many hours did the temperature stay the same?

33. What was the lowest temperature in the nine-hour period?

34. At what time was the temperature over 53°?

35. What was the temperature at 9:00 P.M.?

Make an input-output table for the function. Use 0, 1, 2, and 3 as the domain. (1.8)

36. $y = 2x + 4$

37. $y = 6x + 2$

38. $y = 4x$

39. $y = 20 - 5x$

Review and Assess

ANSWERS

Chapter Support

Parent Guide
Chapter 1

1.1: about 33 mi/h **1.2:** 20, 100 **1.3:** 81 m² **1.4:** 60 is the speed, *t* is the driving time, and 270 is the distance traveled; no; yes **1.5:** $x + 2 = 8$ **1.6:** 6 juices **1.7:** *Sample answer:* from 0 to $30,000 by $5000 **1.8:** 8, 5, 2

Activity Sample Answers

$1 = \frac{4}{4} + (4 - 4)$, $2 = \frac{4}{4} + \frac{4}{4}$, $3 = \frac{(4 + 4 + 4)}{4}$

$4 = 4 + \left(\frac{4 - 4}{4}\right)$, $5 = \frac{(4 \cdot 4 + 4)}{4}$,

$6 = 4 + \left(\frac{4 + 4}{4}\right)$, $7 = 4 + 4 - \left(\frac{4}{4}\right)$,

$8 = 4 + 4 + (4 - 4)$, $9 = 4 + 4 + \left(\frac{4}{4}\right)$,

$10 = \frac{(44 - 4)}{4}$

Prerequisite Skills Review

1. 36 **2.** 8.8 **3.** 0.03 **4.** 1 **5.** 1 **6.** $\frac{1}{4}$

7. $501 < 510$ **8.** $98.02 < 98.021$

9. $7895 > 7885$ **10.** $0.85 > 0.085$

11. $4041 < 4401$ **12.** $0.0022 < 0.02$

13. Area $= 12$ cm²; Perimeter $= 16$ cm

14. Area $= 27.04$ m²; Perimeter $= 20.8$ m

15. Area $= 19.84$ cm²; Perimeter $= 19$ cm

16. Area $= 23.04$ cm²; Perimeter $= 19.2$ cm

Strategies for Reading Mathematics

1. The symbol ÷ means divided by; x^2 means the square of an unknown number; $-$ means minus. **2.** If you look at the step together with the step above it, and compare the numbers, you can find which numbers have changed. Use those numbers to determine the operations that have been performed. For example, in the solution of the example, 16 was the power of 4^2 and 2 is the quotient of 32 and 16. **3.** Order of operations is a way to evaluate expressions that involve more than one operation. First evaluate expressions inside parentheses, then powers, then do multiplications and divisions from left to right, then additions and subtractions from left to right.

4. $3(x^2 + 5) - 1$, for $x = 3$

$= 3(3^2 + 5) - 1$

$= 3(9 + 5) - 1$

$= 3(14) - 1$

$= 42 - 1$

$= 41$

Lesson 1.1

Warm-Up Exercises

1. 21 **2.** $\frac{1}{4}$ **3.** $\frac{2}{3}$ **4.** 400 **5.** 28.4 cm

Lesson Opener

Allow 15 minutes.

1. All answers are 0.

2. Ages will vary. All answers should be 0.

3. Ages will vary. All answers should be 0.

4. The answer will always be 0 no matter what age you start with. *Sample answer:* After adding 5, you multiply by 2, which doubles the age plus 5. So subtracting twice the age and 10 produces 0 no matter what age you start with.

5. Answers will vary. Check work.

Graphing Calculator Activities

1. 3 **2.** 52 **3.** 12 **4.** 25 **5.** 4

6. 7.5 **7.** 12.5 **8.** 16

Practice A

1. subtraction **2.** division **3.** multiplication

4. division **5.** 9 **6.** 15 **7.** 75 **8.** 44

9. 5 **10.** 4 **11.** 9 **12.** 9 **13.** 10

14. 2500 **15.** $\frac{4}{3}$ **16.** 4 **17.** 720 miles

Lesson 1.1 *continued*

18. 2 mi **19.** 288 km **20.** $3x + 4x + 5x$
21. 48 in. **22.** $x + 2x + 30$ **23.** 60 ft

Practice B

1. 12 **2.** 19 **3.** 20 **4.** 44 **5.** 203
6. 4 **7.** 6 **8.** 3 **9.** 32 **10.** 25 **11.** $3\frac{1}{3}$
12. 3 **13.** 0 **14.** 240 **15.** 56
16. 840 mi **17.** 2.5 mi **18.** 196 km
19. $60x + 80x + 100x$ **20.** 192 in.
21. $x + 2x + 60$ **22.** 120 ft **23.** 150 calories
24. 300 calories

Reteaching with Practice

1. 4 minus x, subtraction
2. 7 times y, multiplication **3.** x plus 1, addition
4. y divided by 2, division **5.** 14 **6.** 5
7. 40 **8.** 8 **9.** 108 **10.** 24 **11.** 240 mi
12. 2100 mi **13.** 30 square inches
14. 23 inches

Real-Life Application

1. $I = Prt$
$I = \$1250 \cdot 0.085 \cdot 8$
$I = \$850$

2. \$2100; Answers may vary. *Sample:* Eight years comes from 3 years left of high school and then 5 more years to the reunion.

3. 400 people **4.** 232 people

Challenge: Skills and Applications

1. about 96.54 km/h **2.** 88 ft/sec
3. about 8.32 min **4.** \$268.08

Period	Beginning balance	Interest	Ending balance
1	\$250	\$2.19	\$252.19
2	\$252.19	\$2.21	\$254.40
3	\$254.40	\$2.23	\$256.63
4	\$256.63	\$2.25	\$258.88
5	\$258.88	\$2.27	\$261.15
6	\$261.15	\$2.29	\$263.44
7	\$263.44	\$2.31	\$265.75
8	\$265.75	\$2.33	\$268.08

5. \$17.50; it's \$0.58 less than the \$18.08 from compound interest.

Lesson 1.2

Warm-Up Exercises

1. 16 **2.** 18 **3.** 64 **4.** 9 **5.** 243

Daily Homework Quiz

1. 15 plus x **2.** 45 divided by n **3.** 24 **4.** 3
5. 6 miles

Lesson Opener

Allow 10 minutes.

1. 2; 4; 8; 16; 32; 64; 128
2. 3; 9; 27; 81; 243; 729; 2187
3. 4; 16; 64; 256; 1024; 4096; 16,384
4. 5; 25; 125; 625; 3125; 15,625; 78,125
5. *Sample answer:* The exponent tells you how many times the number is used as a factor. For example, for 6^4 you would take 6 as a factor 4 times, so you would multiply $6 \cdot 6 \cdot 6 \cdot 6$ to get 1296.

Practice A

1. 3^4 **2.** 7^2 **3.** x^3 **4.** y^5 **5.** w^3 **6.** a^3
7. 4^5 **8.** a^2 **9.** 2^3 **10.** x^6 **11.** 5^4
12. $(3x)^2$ **13.** 16 **14.** 1 **15.** 81 **16.** 16
17. 27 **18.** 100,000 **19.** 8 **20.** 100
21. 25 **22.** 5 **23.** 16 **24.** 50 **25.** 100 tiles
26. 384 in.^{2} **27.** 8 ft^3 **28.** 28.26 ft^2

Practice B

1. 3^5 **2.** 9^2 **3.** y^7 **4.** a^4 **5.** g^3
6. $4^5 \cdot b^2$ **7.** a^3 **8.** x^8 **9.** $(3x)^4$ **10.** 36
11. 125 **12.** 144 **13.** 256 **14.** 81
15. 1,000,000 **16.** 216 **17.** 1000 **18.** 25
19. 150 **20.** 15 **21.** 2401 **22.** 561 **23.** 11
24. 36 **25.** 7 **26.** 2744 in.^{3} **27.** about 12.6 ft^2
28. about 7235 in.^{3} **29.** about 87.9 in.^{3}

Reteaching with Practice

1. three to the third power; $3 \cdot 3 \cdot 3$

Lesson 1.2 *continued*

2. five squared; $5 \cdot 5$ **3.** x to the third power; $x \cdot x \cdot x$ **4.** 1000 **5.** 32 **6.** 25 **7.** 1296 **8.** 729 **9.** 243 **10.** 343 **11.** 133 **12.** 20 **13.** 100 **14.** 100 ft^2 **15.** 144 cm^2

Interdisciplinary Application

1. 2500 m^3 **2.** 88,218.88 ft^3
3. 659,877.22 gallons **4.** 1250 m^2
5. 366.4 m^2 **6.** 274.625 ft^2
7. 2054.195 gallons

Challenge: Skills and Applications

1. 64 **2.** 32 **3.** 81 **4.** 4 **5.** 2 **6.** 7 **7.** 4
8. $\frac{1}{16}$ **9.** 16

10.

Figure 4   Figure 5

11.

Figure	1	2	3
Area	5	8	13
Pattern	$1^2 + 4$	$2^2 + 4$	$3^2 + 4$

Figure	4	5
Area	20	29
Pattern	$4^2 + 4$	$5^2 + 4$

12. $n^2 + 4$

Lesson 1.3

Warm-Up Exercises

1. 19 **2.** 27 **3.** 36 **4.** 0 **5.** 100 **6.** 400

Daily Homework Quiz

1. k^2 **2.** 81 **3.** 125 **4.** 144 **5.** 121 m^2

Lesson Opener

Allow 10 minutes.

1. A; The deposit should be added to the number of hours times the cost per hour.

2. B; The amount of tax should be found first. Then the tax can be added to the cost of the jeans and the shipping fee can be added on last.

3. *Sample answer:* If you do not use the correct order, the answer is wrong. In these two cases, you would pay the wrong amount.

Graphing Calculator Activities

1. 13 **2.** 0 **3.** 21 **4.** 8 **5.** 2 **6.** 14

Practice A

1. subtract **2.** multiply **3.** subtract
4. divide **5.** divide **6.** add **7.** 11 **8.** 3
9. 23 **10.** 8000 **11.** 20 **12.** 14 **13.** 8
14. 28 **15.** 82 **16.** 4 **17.** 7 **18.** 5 **19.** 10
20. 14 **21.** 3 **22.** 11 **23.** 8 **24.** 9 **25.** 13
26. 6 **27.** 2 **28.** Calculator 1 **29.** Calculator 1
30. Calculator 2 **31.** Calculator 1
32. $\frac{36 + 35 + 37}{3}$; 36 ft
33. $17 + 17(0.06)$; $18.02
34. $4 \cdot 2 \cdot 3 + 3 \cdot 2 \cdot 1$; 30 ft^3

Practice B

1. $\frac{5}{9}$ **2.** $\frac{40}{3}$ **3.** 1 **4.** 250 **5.** 6 **6.** 8 **7.** 61
8. 28 **9.** 256 **10.** 13 **11.** $\frac{9}{2}$ **12.** 11 **13.** $\frac{9}{4}$
14. 16 **15.** $\frac{2}{5}$ **16.** 23 **17.** 11 **18.** 8 **19.** 9
20. 4 **21.** 2 **22.** 12 **23.** 18 **24.** 0 **25.** 100
26. calculator 1 **27.** calculator 2
28. calculator 2 **29.** calculator 1
30. $\frac{51 + 50 + 58}{3}$; 53 ft
31. $14.95 + 14.95(0.06)$; $15.85

Reteaching with Practice

1. 16 **2.** 8 **3.** 11 **4.** 15 **5.** 8 **6.** 5
7. 3 **8.** 36 **9.** 9 **10.** $157

Real-Life Application

1. *Sample Answer:*
$[2(18 - 4.5) + 2(29.95 - 5.99) +$
$4(9.95 - 1.99) + 2(32 - 6.4)$
$+ 2(65.95 - 26.38) + 2(12.95 - 2.59)$
$+ 4(9) + 2(25)] \div 2$
2. $171.91

Lesson 1.3 *continued*

3. *Sample answer:*
$2(18.00 + 29.95 + 32.00 + 65.95 + 12.95 + 25.00) + 4(9.95 + 9.00) - 2(171.91)$
4. $99.68 **5.** $28.09 (Writing answers will vary. Total amount must be less than $28.09.)

Challenge: Skills and Applications

1. 1.3 **2.** 7.7 **3.** $27.69, $33.23, $37.98, $17.41 **4–7.** Accept equivalent forms.

4. *Sample answers:* $0.75p + 0.055(0.75p)$, or $1.055(p - 0.25p)$ **5.** *Sample answer:* $(a^2 - b^2)h$
6. *Sample answer:* $(ab + a^2)h$ **7.** *Sample answer:* $\left(\frac{1}{2} ab + b^2\right)h$

Quiz 1

1. 20 **2.** 4 **3.** 90 mi **4.** 625 **5.** 49
6. 216 ft^3 **7.** 18 **8.** 333 **9.** 44

Lesson 1.4

Warm-Up Exercises

1. 2 **2.** 192 **3.** 24 **4. a.** false **b.** true

Daily Homework Quiz

1. 20 **2.** 3 **3.** 84 **4.** 25
5. $75 \cdot 15 - 200$; $925

Lesson Opener

Allow 10 minutes.

1. a. *Sample answer:* Subtract 2 from 36 to get 34. **b.** If you replace x with 34, you get 36.

2. a. *Sample answer:* Multiply 4 by 3 to get 12. **b.** If you replace x with 12, you get 4.

3. a. *Sample answer:* Subtract 5 from 29 to get 24. Then divide 24 by 8 to get 3. **b.** If you replace x with 3, you get 29.

Practice A

1. equation **2.** expression **3.** inequality
4. equation **5.** inequality **6.** expression
7. no **8.** no **9.** yes **10.** no **11.** yes
12. no **13.** yes **14.** no **15.** yes **16.** no
17. yes **18.** yes **19.** What number minus 5 gives 3?; 8 **20.** What number plus 2 gives 6?; 4

21. What number added to 4 gives 6?; 2
22. What number subtracted from 12 gives 5?; 7
23. What number times 4 gives 20?; 5
24. What number divided into 30 gives 6?; 5
25. What number divided by 2 gives 3?; 6
26. What number divided into 18 gives 3?; 6
27. What number cubed gives 8?; 2
28. yes **29.** no **30.** no **31.** no **32.** yes
33. yes **34.** yes **35.** yes **36.** yes
37. 8 is the width in inches of a new locker; x is the number of new lockers that can be installed; 144 is the length in inches of the space available for new lockers; 18 lockers.

38. 14 is the number of lamps; x is the number of watts in 1 lamp; 14,000 is the total watts; 1000 watts.

Practice B

1. yes **2.** no **3.** no **4.** yes **5.** yes **6.** no
7. yes **8.** yes **9.** no **10.** no **11.** yes
12. no **13.** What number minus 5 gives 8?; 13
14. What number plus 2 gives 12?; 10
15. Nine less than 6 times what number gives 9?; 3
16. What number subtracted from 24 gives 15?; 9
17. What number times 4 gives 32?; 8
18. Seven more than 3 times what number gives 22?; 5
19. What number divided by 7 gives 3?; 21
20. What number divided into 36 gives 3?; 12
21. What number cubed gives 64?; 4
22. yes **23.** no **24.** no **25.** yes **26.** yes
27. yes **28.** no **29.** yes **30.** no
31. B **32.** D **33.** E **34.** A **35.** C
36. 8 is the width in inches of a new locker; x is the number of new lockers that can be installed; 200 is the length in inches of the space available for new lockers; 25 lockers. **37.** No; 5 is the savings per week; n is the number of weeks; 145 is the total savings needed; at least 29 weeks.

Reteaching with Practice

1. 3 is not a solution. **2.** 3 is not a solution.
3. 2 is not a solution. **4.** 12 is a solution.
5. 5 is not a solution. **6.** 3 is a solution.

Lesson 1.4 *continued*

7. What number plus 7 gives 21? The number is 14. **8.** One more than 3 times what number gives 19? The number is 6. **9.** What number minus 12 gives 10? The number is 22.
10. What number divided by 3 gives 11? The number is 33. **11.** Seven less than 4 times what number gives 9? The number is 4. **12.** What number divided by 2 gives 4? The number is 8.
13. 5 is not a solution. **14.** 3 is not a solution.
15. 2 is a solution.

Interdisciplinary Application

1. $12h \geq 200$ **2.** Twelve is the number of weeks available in the summer, h is the number of hours spent practicing each week, and 200 is the total hours spent practicing. **3.** 17 hours or more **4.** Answers may vary.

Challenge: Skills and Applications

1. no **2.** no **3.** yes **4.** yes **5.** yes **6.** no

7. no **8.** yes **9.** $50,000 = \dfrac{0.015(4000)^2}{f}$

10. no **11.** no **12.** Between 4 and 5 Newtons

Lesson 1.5

Warm-Up Exercises

1. C **2.** A **3.** B

Daily Homework Quiz

1. not a solution **2.** is a solution **3.** $n = 22$
4. $y = 12$ **5.** 5 days

Lesson Opener

Allow 10 minutes.

1. C; You add the cost of all three pieces together to get the total cost of the system.

2. A; Since x represents the amount you spent on each of the last two days, you multiply x by 2 and add 21 to get the total amount spent.

3. C; Since x represents the number of three-point baskets scored, x times 3 plus the other points scored should be 56.

Practice A

1. $x + 3$ **2.** $x - 4$ **3.** $8 - x$ **4.** $x + 1$

5. $6x$ **6.** $\dfrac{1}{2}x$ **7.** $\dfrac{x}{5}$ **8.** $2x + 7$ **9.** $\dfrac{x - 2}{9}$

10. $10x + 2$ **11.** $3(x + 1)$ **12.** $\dfrac{x + 6}{2}$

13. $x + 2 = 10$ **14.** $y + 4 = 13$
15. $y + 8 \geq 9$ **16.** $a - 2 = 7$
17. $z - 6 < 15$ **18.** $11 - b = 2$
19. $2x = 22$ **20.** $12 < 6x$ **21.** $4b + 1 = 5$

22. $\dfrac{t}{3} = 8$ **23.** $\dfrac{a}{2} > 5$ **24.** $6a - 4 = 8$

Practice B

1. $x + 4$ **2.** $x - 6$ **3.** $12 - x$ **4.** $x + 2$

5. $5x$ **6.** $\dfrac{1}{3}x$ **7.** $\dfrac{x}{8}$ **8.** $2x + 9$ **9.** $\dfrac{x - 2}{3}$

10. $10x + 3$ **11.** $5(x + 1)$ **12.** $\dfrac{x + 5}{2}$

13. $x + 7 = 10$ **14.** $y + 6 = 13$
15. $y + 8 \geq 10$ **16.** $a - 2 = 8$
17. $z - 6 < 21$ **18.** $13 - b = 2$
19. $11x = 22$ **20.** $14 < 7x$ **21.** $4b + 1 = 17$

22. $\dfrac{t}{3} = 9$ **23.** $\dfrac{a}{2} > 9$ **24.** $6a - 3 = 9$

25. $x + 7 = 13; 6$ **26.** $25 - x = 15; 10$

27. $6x = 54; 9$ **28.** $\dfrac{x}{8} = 9; 72$

Reteaching with Practice

1. $x + 1$ **2.** $x - 4$ **3.** $12 - x$ **4.** $x + 8$

5. $x + 6$ **6.** $x - 10$ **7.** $3n$ **8.** $\dfrac{n}{4}$

9. $12n$ **10.** $\dfrac{1}{2}n$ **11.** $\dfrac{9}{n}$ **12.** $x + 6 = 15$;

$x = 9$ **13.** $\dfrac{x}{4} = 7; x = 28$

14. $x + 8 = 15; x = 7$

15. $x \cdot 3 = 33; x = 11$

Real-Life Application

1. 6 **2.** 9 **3.** NT\$1750 saved **4.** \$255.63

Challenge: Skills and Application

1. $w + 2$ **2.** $(w)(w + 2) = 120$ ft^2

3. 10 ft $\times 12$ ft   **4.** $l - 3$

5. $(l)(l - 3) = 180$ ft^2 **6.** 12 ft $\times 15$ ft

7. $\dfrac{3d}{2}$ **8.** $(d)\left(\dfrac{3d}{2}\right) = 216$ ft^2

9. 12 ft $\times 18$ ft   **10.** $283.90   **11.** $(t - 7)°$

12. $(t + 12)°$ **13.** $53°$

Quiz 2

1. no; $24 \neq 16$ **2.** yes; $72 \geq 65$

3. *Sample answer*: The product of what number and 3 can be added to 2 to get 14?   **4.** $4x - 7$

5. $h + 20 < 35$ **6.** $20s = 600$

Lesson 1.6

Warm-Up Exercises

1. 2   **2.** 10   **3.** 6   **4.** 2   **5.** 9

Daily Homework Quiz

1. $n - 80$ **2.** $27n$ **3.** $y + 16 < 35$

4. $\dfrac{y}{50} = 3$ **5.** 4 hours

Lesson Opener

1. B   **2.** A

Practice A

1. a   **2.** b   **3.** b   **4.** b

5. Sample answer:
  s speed of airplane
  2   flight time
  360 distance traveled

6. $s \cdot 2 = 360$

7. 180 miles per hour

8. $180 \dfrac{\text{miles}}{\text{hour}} \cdot 2 \text{ hours} = 360 \text{ miles}$

Practice B

1. a   **2.** a   **3.** b   **4.** a   **5.** b

6.

Money paid per can	·	Number of cans recycled	=	Total money paid

7. *Sample answer:*
  m money paid per can
  n number of cans recycled
  5 total money paid
  $0.1 \cdot n = 5$

8. 50 cans; $\dfrac{\$.10}{\text{can}} \cdot 50 \text{ cans} = \$5.00$

Reteaching with Practice

1. a.

Price per hat	·	Number of hats	=	Total receipts

b. Price per hat = 8   (dollars)
  Number of hats = n (hats)
  Total receipts = 2480   (dollars)
    $8n = 2480$

c. 310 hats

2. a.

Cost per person	·	Number of persons	=	Total cost

b. Cost per person = 7.50   (dollars)
  Number of persons = 3   (persons)
  Total cost = n (dollars)
    $7.50(3) = n$

c. $22.50

3. $\dfrac{2}{15}$ h or 8 min

Real-Life Application

1. $f = \$.10(m - 4y)$ **2.** $586.50

3. $t = d + p + 48i$

4 a. Plan A: $46,612 Plan B: $46,132
  Plan C: $45,172 Plan D: $44,212

b. Plan A   **c.** Plan D   **d.** $5712

5. 69,600 miles; $(15,000 \cdot 4) + \dfrac{(46,132 - 45,172)}{\$.10}$

Challenge: Skills and Applications

1. $72

2. $\dfrac{\text{money saved} + \text{money need to save}}{\text{number of weeks}} =$
average savings

Lesson 1.6 *continued*

3. money saved = 72; money need to save = x; number of weeks = 5; average savings = 20

4. $\dfrac{72 + x}{5} = 20$ **5.** 28 **6.** I need to save $28 this week. **7.** *Sample answer:* $1.0575(9.5x) \le 48$ **8.** 4 **9.** 4 CDs **10.** no **11.** yes

12. *Sample answer:* $65t \ge 143$

Lesson 1.7

Warm-Up Exercises
1. 322,700,000 **2.** 370,000,000

Daily Homework Quiz
1. Number of friends · Each share = Bill + Tax and tip

2. Number of friends = 8
Each share = d
Bill = 130
Tax and tip = 30

3. $8d = 130 + 30$

4. $20

Lesson Opener
Allow 10 minutes.

1. Arachnids; 5 is less than the other numbers listed. **2.** Birds; 75 is greater than the other numbers listed. **3.** harder; The number of species would not be listed in numerical order.

4, 5. Accept reasonable estimates.

4. 1995; about 66% **5.** 1980; about 23%

Practice A
1. 12th grade **2.** 10th and 11th **3.** bus
4. 1985 **5.** 15 **6.** secondary **7.** 27 million
8. 1985 **9.**

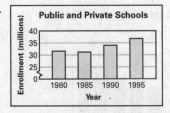

10. August **11.** $20,000

12.

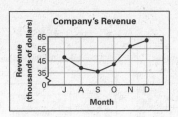

13. 1960s **14.** 1940s

15. *Sample answer:* Life expectancy has increased over the 70-year period, and the greatest increase was during the 1940s.

Practice B
1. 10th grade **2.** 10th and 11th **3.** bus
4. 1985 **5.** 19 **6.** secondary **7.** false
8. true **9.**

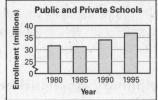

10. November

11. $11,000

12.

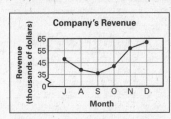

13. No; expenses exceeded revenue by $2000 so it was not a profitable six months.

14. 1940s

15.

Sample Answer: Life expectancy has increased over the 70-year period, and the greatest increase was during the 1940s.

Reteaching with Practice
1. 1970 **2.** 1995

Lesson 1.7 *continued*

3. 1985–1990 **4.** 1980–1985

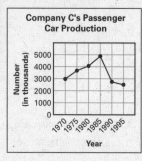

Company C's Passenger Car Production

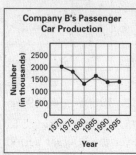

Company B's Passenger Car Production

Learning Activity

1. *Sample answer:* Newspapers use circle graphs because they are easy to read and understand. Other graphs used by the media include bar and line graphs, and pictographs. **2.** *Sample answer:* Circle graphs and pictographs are easy to read, but line and bar graphs are able to more accurately compare differences in data.

Interdisciplinary Application

1.

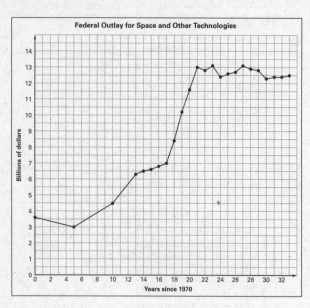

Federal Outlay for Space and Other Technologies

2. 1975–1991 (16 years)

1986–1991 from \$6.8 billion to \$13.0 billion or \$6.2 billion

3. General decrease or static change

4. Answers will vary. *Sample:* More money was spent during the height of the "space race."

Challenge: Skills and Applications

1. Joe 50.5%, Tina 55.75%, Ali 49%, Keesha 51.75%

2.

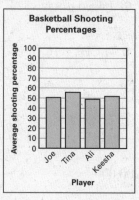

Basketball Shooting Percentages

3.

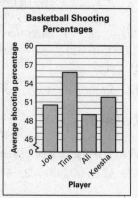

Basketball Shooting Percentages

4. All 4 players have similar shooting averages; Tina's shooting average is a lot better than the other averages.

5. *Sample graph:*

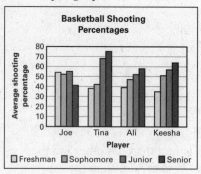

Basketball Shooting Percentages

6. Joe **7.** Tina **8.** Keesha **9.** sophomore and junior **10.** junior and senior

Lesson 1.8

Warm-Up Exercises

1. 12 **2.** 15 **3.** 18 **4.** 21

5. *Sample answer:* each increases by 3

Daily Homework Quiz

a.

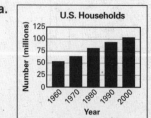

U.S. Households

b. 1970 to 1980

Lesson 1.8 *continued*

Lesson Opener
Allow 10 minutes.

1. 3; 4; 5; 6 **2.** 2; 3; 4; 5 **3.** 4; 5; 6; 7

Practice A

1. function **2.** input **3.** output

4. output; input

5. Yes; For each input, there is exactly one output. **6.** Yes; For each input, there is exactly one output. **7.** No; There are two output values that correspond to each input value 1 and 2.

8. Yes; For each input, there is exactly one output. **9.** No; There are two output values that correspond to the input value of 4. **10.** Yes; For each input, there is exactly one output.

11.

Input	Output
0	3
1	5
2	7
3	9

12.

Input	Output
0	0
1	3
2	6
3	9

13.

Input	Output
0	9
1	8
2	7
3	6

14.

Input	Output
0	5
1	6
2	7
3	8

15.

Input	Output
0	$\frac{1}{2}$
1	$1\frac{1}{2}$
2	$2\frac{1}{2}$
3	$3\frac{1}{2}$

16.

Input	Output
0	10
1	7
2	4
3	1

17. $p = 3n + 10$ **18.**

Input	Output
1	13
2	16
5	25
8	34
10	40

19.

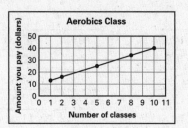

20. $d = 80t$ **21.**

Input	Output
2	160
5	400
8	640
10	800

Practice B

1. function **2.** input; output **3.** output; input

4. input-output table **5.** Yes; For each input, there is exactly one output. **6.** Yes; For each input, there is exactly one output. **7.** No; There are two output values that correspond to the input value 2.

8.

Input	Output
0	5
1	7
2	9
3	11

9.

Input	Output
0	0
1	4
2	8
3	12

10.

Input	Output
0	15
1	14
2	13
3	12

11.

Input	Output
0	18
1	19
2	20
3	21

12.

Input	Output
0	$\frac{1}{2}$
1	$2\frac{1}{2}$
2	$4\frac{1}{2}$
3	$6\frac{1}{2}$

13.

Input	Output
0	20
1	17
2	14
3	11

Lesson 1.8 *continued*

14.

Input	Output
2	7.5
2.5	9
4	13.5
5	16.5
5.5	18

15.

Input	Output
2	18
2.5	15.8
4	12.5
5	11.4
5.5	11

16.

Input	Output
2	1.5
2.5	3.75
4	13.5
5	22.5
5.5	27.75

17. $p = 3n + 10$

18.

Input	Output
1	13
2	16
5	25
8	34
10	40

19.

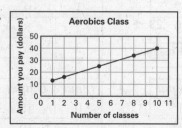

20. domain: $\{1, 2, 3, \ldots\}$,
range: $\{13, 16, 19, \ldots\}$

21. $d = 80t$

22.

Input	Output
2	160
5	400
8	640
10	800

23.

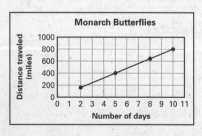

Reteaching with Practice

1. The table represents a function. For each input, there is exactly one output.

2. The table does not represent a function. For the input value 2, there are two output values, not one.

3.

Input x	Output y
0	5
1	4
2	3
3	2

4.

Input x	Output y
0	1
1	5
2	9
3	13

5.

Input x	Output y
0	9
1	8
2	7
3	6

6. $C = 1.5 + 4p$

7. $C = 3 + 2.50p$

Cooperative Learning Activities

1. As the radius increases one unit, the circumference increases.   **2.** As the radius increases one unit, the area increases (at a higher rate than that of circumference).

Interdisciplinary Application

1. $D = 7920 + 1248t$

2.

Time (minutes)	35	40	45
Distance (feet)	51,600	57,840	64,080

Time (minutes)	50	55
Distance (feet)	70,320	76,560

3. 14.5 miles

4.

Input (t)	20	40	60
Output (feet)	8800	17,600	26,400

Input (t)	80	100	120
Output (feet)	35,200	44,000	52,800

Input (t)	140	160	180
Output (feet)	61,600	70,400	79,200

Answer: between 160 and 180 min

Answers

Lesson 1.8 *continued*

Challenge: Skills and Applications

1.

Input	Output
1	4.9
1.25	6.65
2.35	14.35
2.8	17.5
3	18.9

2.

Input	Output
1	9.4
1.25	9.45
2.35	9.67
2.8	9.76
3	9.8

3.

Input	Output
1	0.3
1.25	0.8625
2.35	4.8225
2.8	7.14
3	8.3

4.

Input	Output
1	0.18
1.25	0.36125
2.35	1.90125
2.8	2.88
3	3.38

5. function: there is only one amount of sales tax for each purchase amount **6.** not a function; the change could be an increase or a decrease, making two possible temperatures resulting from one initial temperature **7.** function: for each year there is only one number of days

8. range: 365 and 366 **9.** $y = x^2 + 3$

10. $y = \dfrac{12}{x}$

Review and Assessment

Review Games and Activities

Across 1. quotient **3.** exponent

5. inequality **6.** solution **8.** verbal

9. expression **13.** grouping symbols

15. variable **18.** difference **19.** squared

20. cubed

Down 2. triangles **3.** equation

4. power **7.** function **10.** range **11.** sum

12. domain **14.** product **16.** base **17.** data

Test A

1. 3 **2.** 136 **3.** 420 mi **4.** 18*a* **5.** 64

6. 125 **7.** 243 **8.** 216 **9.** 19 **10.** 512

11. 684 cm³ **12.** 6 **13.** 5 **14.** no **15.** yes

16. yes **17.** no **18.** 8 + *x* **19.** *x* − 10

20. 3 + *x* = 10 **21.** 4 > 6*t* **22.** $d = \frac{3}{2} + p$

23. 4*s* ≥ 70 + 5 **24.** *k* ≥ *a* + 3

25.

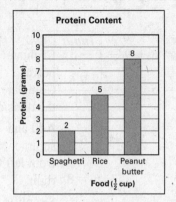

Protein Content

26.

Input	Output
0	2
1	9
2	16
3	23

27.

Input	Output
0	12
1	9
2	6
3	3

Test B

1. 150 **2.** 8 **3.** 260 mi **4.** 56*x*²

5. 64 **6.** 256 **7.** 253 **8.** 56 **9.** 86

10. 256 **11.** 9680 in.³ **12.** 25 **13.** 108

14. 18 **15.** 0 **16.** no **17.** no **18.** $x + \frac{1}{2}$

19. $\frac{2}{3}x$ **20.** 3 ≤ 4 − *x* **21.** 2³ = 8

22. 400 ≤ 32*p* **23.** 8*m* + 8 · 5 = 6 · 60

24.

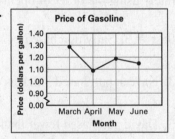

Price of Gasoline

25.

Input	Output
1	5.1
1.5	7.6
2	11.1
3	21.1

26.

Input	Output
1	5.5
1.5	3.5
2	2.5
3	1.5

SAT/ACT Chapter Test

1. C **2.** A **3.** B **4.** B **5.** C **6.** A

7. A **8.** C **9.** B **10.** C

Alternative Assessment

1. a–c. Complete answers should include these points.

a.

$$6 + 3^2 \div 3 - 5 = 6 + 9 \div 3 - 5 \qquad \text{Evaluate power.}$$
$$= 6 + 3 - 5 \qquad \text{Complete the division.}$$
$$= 9 - 5 \qquad \text{Add and subtract from left to right.}$$
$$= 4 \qquad \text{Simplify.}$$

Amit is correct, since he used the order of operations.

b. • Explain order of operations: First do operations that occur within grouping symbols. Then evaluate powers. Multiply and divide from left to right. Finally, add and subtract from left to right.

• Explain that it is important to have an established order of operations for there to be consistency within mathematics.

c. • Explain that the calculator was not programmed with the established order of operations and completed all calculations from left to right.

• Explain that Sue should be sure to enter each computation separately using the order of operations herself.

2. a. $1 + 3x = 10$; $x = 3$, Sally purchased exactly 3 tickets.

$1 + 3x \le 10$; $x \le 3$, Naveen purchased 3 tickets or fewer.

$1 + 3x < 10$; $x < 3$, David purchased fewer than 3 tickets.

b. 3 rides for Sally and Naveen, 2 rides for David

3. a. The number of rides for both can be less than 3, but for Naveen it can also be 3. The key words are *less than $10* or *$10 or less* because they change the meaning of the inequality.

b. No; Naveen may take 3 or fewer rides; Sally must take 3.

4. Answers may vary. *Sample answer:* Rita can spend $10 or more.

Project: Watch It Disappear

1. Student tables should include three rows: day, water level, and change from previous day. The entry in the change from previous day row should be the difference between two consecutive days.

2. Graphs should show the water level decreasing, but not necessarily at a constant rate.

3. Yes; for each time there is only one water level. **4.** Students should add the changes in the table and divide by 4.

5. Original level − (average change per day) × (day) = water level on that day. *Sample answer:* $y = 1 - \frac{1}{8}x$.

6. Students should substitute the day number for x in the equation in Exercise 5 and find y.

7. *Sample answer:* $1 - \frac{1}{8}x = 0$; 8 days

Cumulative Review

1. 10 **2.** 12 **3.** 15 **4.** 1 **5.** 30 **6.** 4

7. 25 **8.** 1000 **9.** 1225 **10.** 25 **11.** 10

12. 225 **13.** 3 **14.** 7 **15.** $\frac{1}{8}$ **16.** 16

17. Yes, 3 is a solution. **18.** Yes, 2 is a solution.

19. No, 4 is not a solution.

20. Yes, 6 is a solution. **21.** $9n$ **22.** $x > 20$

23. $\frac{a}{5}$ **24.** $ab < 10$ **25.** 9 months **26.** 6 CDs

27. 162 calories **28.** Domain: 0, 2, 4, 6, 8, 10; Range: 7, 10, 12, 15, 17, 21 **29.** Domain: 0, 10, 20, 30, 40, 50; Range: 21, 25, 28, 32, 35, 43

30.

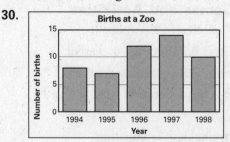

31.

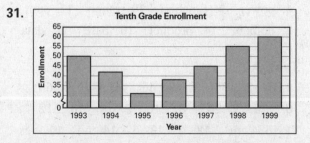

32. 1 hour **33.** 40° **34.** 5:00 P.M. **35.** 50°

Review and Assessment *continued*

36.

Input	Output
0	4
1	6
2	8
3	10

37.

Input	Output
0	2
1	8
2	14
3	20

38.

Input	Output
0	0
1	4
2	8
3	12

39.

Input	Output
0	20
1	15
2	10
3	5